SpringerWienNewYork

CISM COURSES AND LECTURES

The series presents lecture notes, monographs, edited works and proceedings in the field of Mechanics, Engineering, Computer Science and Applied Mathematics.
Purpose of the series is to make known in the international scientific and technical community results obtained in some of the activities organized by CISM, the International Centre for Mechanical Sciences.

INTERNATIONAL CENTRE FOR MECHANICAL SCIENCES

COURSES AND LECTURES - No. 529

DYNAMICAL INVERSE PROBLEMS: THEORY AND APPLICATION

EDITED BY

GRAHAM M.L. GLADWELL
UNIVERSITY OF WATERLOO, ONTARIO, CANADA

ANTONINO MORASSI
UNIVERSITY OF UDINE, ITALY

SpringerWienNewYork

This volume contains 83 illustrations

SPIN 80035948

All contributions have been typeset by the authors.

ISBN 978-3-7091-0695-2 SpringerWienNewYork

PREFACE

Classical vibration theory is concerned with the determination of the response of a given dynamical system to a prescribed input. These are called direct problems in vibration and powerful analytical and numerical methods are available nowadays for their solution. However, when one studies a phenomenon which is governed by the equations of classical dynamics, the application of the model to real life situations often requires the knowledge of constitutive and/or geometrical parameters which in the direct formulation are considered as part of the data, whereas, in practice, they are not completely known or are inaccessible to direct measurements. Therefore, in several areas in applied science and technology, one has to deal with inverse problems in vibration, that is problems in which the roles of the unknowns and the data is reversed, at least in part. For example, one of the basic problems in the direct vibration theory - for infinitesimal undamped free vibrations - is the determination of the natural frequencies and normal modes of the vibrating body, assuming that the stiffness and mass coefficients are known. In the context of inverse theory, on the contrary, one is dealing with the construction of a model of a given type (i.e., a mass-spring system, a string, a beam) that has given eigenproperties.

In addition to its applications, the study of inverse problems in vibration has also inherent mathematical interest, since the issues encountered have remarkable features in terms of originality and technical difficulty, when compared with the classical problems of direct vibration theory. In fact, inverse problems do not usually satisfy the Hadamard postulates of well-posedeness, also, in many cases, they are extremely non-linear, even if the direct problem is linear. In most cases, in order to overcome such obstacles, it is impossible to invoke all-purpose, ready made, theoretical procedures. Instead, it is necessary to single out a suitable approach and trade-off with the intrinsic ill-posedeness by using original ideas and a deep use of mathematical methods from various areas. Another specific and fundamental aspect of the study of inverse problems in vibration concerns the numerical treatment and the development of ad-hoc strategies for the treatment of ill-conditioned, linear and non-linear problems. Finally, when inverse techniques are applied to the study of real problems, additional obstructions arise because of the complexity of mechanical modelling, the inadequacy of the analytical models used for the interpretation of the experiments, measurement errors and incompleteness of the field data. Therefore, of particular relevance for practical applications is to assess the robustness of the algorithms to measurement errors and to the accuracy of the analytical models used to describe the physical phenomenon.

The purpose of the CISM course entitled "Dynamical Inverse Problems: Theory and Application", held in Udine on May 25-29 2009, was to present a state-of-the-art overview of the general aspects and practical applications of dynamic inverse methods, through the interaction of several topics, ranging from classical and advanced inverse problems in vibration, isospectral systems, dynamic methods for structural identification, active vibration control and damage detection, imaging shear stiffness in biological tissues, wave propagation, computational and experimental aspects relevant for engineering problems.

The course was addressed to PhD students and researchers in civil and mechanic engineering, applied mathematics, academic and industrial researchers.

In the first chapter Gladwell discusses matrix inverse eigenvalue problems. He describes the classical inverse problems for in-line systems, such as discrete models of rods in longitudinal vibration and beams in flexural vibration. He illustrates the theory governing the formation of isospectral systems, and describe how it may be used to construct isospectral finite-element models of membranes. Throughout the chapter, emphasis is placed on ways of choosing data that lead to a realistic system. Morassi in the second chapter describes some classical approaches to the inversion of continuous second-order systems. Attention is focused on uniqueness, stability and existence results for Sturm-Liouville differential operators given in canonical form on a finite interval. A uniqueness result for a fourth order Euler-Bernoulli operator of the bending vibration of a beam is also discussed. The next chapter by Röhrl presents a method to numerically solve the Sturm-Liouville inverse problem using a least squares approach based on eigenvalue data. The potential and the boundary conditions are estimated from two sequences of spectral data in several examples. Theorems show why this approach works particularly well. An introduction to the Boundary Control method (BC-method) for solving inverse problems is presented by Belishev in the fourth chapter. In particular, the one-dimensional version of the BC-method is used for two dynamical inverse problems. The first problem is to recover the potential in a Sturm-Liouville operator describing the transverse vibration of a semi-infinite taut string with constant linear mass density by time-history measurements at the endpoint of the string. The second problem deals with a second-order vectorial dynamical system governing, for example, the longitudinal vibration of two semi-infinite connected beams having constant linear mass densities. An inverse problem is to recover the matrix coefficients of the lower order terms via time-domain measurements at the endpoint of the beam. Connections between the BC-method and asymptotic methods in PDEs, functional analysis, control and system theory, are especially investigated in this chapter. In the fifth chapter Vestroni and Pau introduce dynamic methods for dynamic characterization and damage identification of civil engineering structures. Indeterminacy and difficulties in modal identification and model updating are discussed with reference to several experimental cases of masonry structures. A damage identification procedure in a steel arch with concentrate damage is also presented. Eigenvalue assignment problems in vibration are presented by Mottershead, Tehrani and Ram in the sixth chapter. Procedures are described for pole placement by passive modification and active control using measured receptances. The theoretical basis of the method is described and experimental implementation is explained. The book ends with the lectures by Oberai and Barbone on inverse problems in biomechanical imaging. The authors briefly describe the clinical relevance of these problems and how the measured data is acquired. Attention is focused on two distinct strategies for solving these problems. These include a direct approach that is fast but relies on the availability of complete interior displacement measurements. The other is an iterative approach that is computationally intensive but is able to handle incomplete interior data and is robust to noise.

Graham M.L. Gladwell and Antonino Morassi

CONTENTS

Matrix Inverse Eigenvalue Problems

Graham M.L. Gladwell
Department of Civil Engineering,
University of Waterloo, Waterloo, Ontario, Canada

Abstract The term matrix eigenvalue problems refers to the computation of the eigenvalues of a symmetric matrix. By contrast, the term inverse matrix eigenvalue problem refers to the construction of a symmetric matrix from its eigenvalues. While matrix eigenvalue problems are well posed, inverse matrix eigenvalue problems are ill posed: there is an infinite family of symmetric matrices with given eigenvalues. This means that either some extra constraints must be imposed on the matrix, or some extra information must be supplied. These lectures cover four main areas: i) Classical inverse problems relating to the construction of a tridiagonal matrix from its eigenvalues and the first (or last) components of its eigenvectors. ii) Application of these results to the construction of simple in-line mass-spring systems, and a discussion of extensions of these results to systems with tree structure. iii) Isospectral systems (systems that all have the same eigenvalues) studied in the context of the QR algorithm, with special attention paid to the important concept of total positivity. iv) Introduction to the concept of Toda flow, a particular isospectral flow.

1 Lecture 1. Classical Inverse Problems

1.1 Introduction

The word *inverse*, the related terms *reverse*, and *mirror image*, implicitly state that there is some 'object' that is related to another 'object' called its inverse and just as a physical object and its image in a mirror are related. Then talking about mirrors, we tend to label one object 'real' and the other 'the reflection' suggesting that the 'reflection' is not 'real'. In our situation, dealing with mathematical objects that are 'problems', again we often label one problem as the 'real' one and the other, the inverse, as somehow not so real. Strictly speaking, we will just have two problems and can label one a 'direct' problem and the other the 'inverse' problem. Which problem is called direct and which inverse is often an accident of history: the direct one is the familiar one.

The inverse problems we will analyze are all variants of so-called *matrix inverse eigenvalue problems.* For an exhaustive catalogue of such problems, see Chu and Golub (2005). See also Xu (1998) and Gladwell (2004). Because we are interested primarily in problems related to the undamped vibrations of elastic structures, we limit our attention to eigenvalue problems related to a real, square, symmetric matrix. We denote the set of real square matrices of order n by M_n, the subset of symmetric matrices by S_n, and the set of real $n \times 1$ column vectors by R_n.

If $\mathbf{A} \in S_n$, then a value of λ for which there is a non-zero $\mathbf{x} \in R_n$ such that

$$(\mathbf{A} - \lambda\mathbf{I})\mathbf{x} = \mathbf{0} \tag{1}$$

is called an *eigenvalue* of $\mathbf{A}$; $\mathbf{x}$ is called an *eigenvector* of $\mathbf{A}$ corresponding to λ. It is known that $\mathbf{A} \in S_n$ has exactly n real eigenvalues $(\lambda_i)_1^n$, not necessarily distinct, and n corresponding eigenvectors $(\mathbf{x}_i)_1^n$ that span the space R_n. If $\lambda_i \neq \lambda_j$, then $\mathbf{x}_i$ and $\mathbf{x}_j$ are orthogonal: $\mathbf{x}_i^T\mathbf{x}_j = \mathbf{0}$; if λ_i is an r-fold multiple eigenvalue, then we can construct r mutually orthogonal eigenvectors corresponding to λ_i. The eigenvectors $(\mathbf{x}_i)_1^n$ can be scaled so that they are unit vectors, and then they are *orthonormal*:

$$\mathbf{x}_i^T\mathbf{x}_j = \delta_{ij} = \begin{cases} 1 & i = j \\ 0 & i \neq j \end{cases} .$$

The matrix $\mathbf{X} \in M_n$ with columns $(\mathbf{x}_i)_1^n$ is an orthogonal matrix: it satisfies

$$\mathbf{X}^T\mathbf{X} = \mathbf{X}\mathbf{X}^T = \mathbf{I}.$$

The set of orthogonal matrices in M_n is denoted by O_n.

The direct eigenvalue problem for $\mathbf{A} \in S_n$ is this: *Find the eigenvalues of* $\mathbf{A}$. It can be shown that this is a *well-posed problem.*

The definition of a *well-posed problem* is due to Jacques Hadamard. His 1902 definition has three elements:

Existence: the problem has a solution
Uniqueness: the problem has no more than one solution
Continuity: the solution is a continuous function of the data

For the direct problem, *find the eigenvalues of* $\mathbf{A}$, we know that $\mathbf{A}$ *has* a *unique* set of eigenvalues, so that Properties 1 and 2 are satisfied. Since the eigenvalues appear as roots of a polynomial equation whose coefficients are additive and/or multiplicative combinations of the entries of $\mathbf{A}$, they will be continuous functions of these entries: Property 3 holds; the problem is well-posed.

If $\mathbf{Q} \in O_n$ and $\mathbf{A} \in S_n$, then $\mathbf{B} = \mathbf{Q}^T\mathbf{A}\mathbf{Q}$ will have the same eigenvalues as $\mathbf{A}$: if (1) holds, then

$$\mathbf{Q}^T(\mathbf{A} - \lambda\mathbf{I})\mathbf{x} = \mathbf{0}.$$

But $\mathbf{Q}^T\mathbf{A}\mathbf{x} = \mathbf{Q}^T\mathbf{A}(\mathbf{Q}\mathbf{Q}^T)\mathbf{x} = \mathbf{B}(\mathbf{Q}^T\mathbf{x}) = \mathbf{B}\mathbf{y}$, where $\mathbf{y} = \mathbf{Q}^T\mathbf{x}$ and thus

$$(\mathbf{B} - \lambda\mathbf{I})\mathbf{y} = \mathbf{0}.$$

We can also show that all $\mathbf{B} \in S_n$ with eigenvalues $(\lambda_i)_1^n$ have the form $\mathbf{B} = \mathbf{Q}^T\mathbf{A}\mathbf{Q}$, where $\mathbf{A}$ is one such matrix; we can always take $\mathbf{A} = diag(\lambda_1, \lambda_2, \ldots, \lambda_n) = \Lambda$. Write $N = n(n+1)/2$. The orthogonal matrices $\mathbf{Q}$ form an $(N - n) = n(n-1)/2$-parameter family: the inverse problem '*Find* $\mathbf{A} \in S_n$ *with eigenvalues* $(\lambda_i)_1^n$', is not well-posed. We note that $\mathbf{A} \in S_n$ has N independent entries, and there are only n pieces of data $(\lambda_i)_1^n$: there is an $(N - n)$-parameter family of solutions.

1.2 The Rayleigh Quotient

If $\mathbf{x} \in R_n$ and $\mathbf{A} \in S_n$, the ratio

$$R = \frac{\mathbf{x}^T\mathbf{A}\mathbf{x}}{\mathbf{x}^T\mathbf{x}}$$

is called the *Rayleigh Quotient*, after Rayleigh (1894). The primary result relating to the Rayleigh Quotient is that *the stationary values of* R *for varying vectors* $\mathbf{x}$ *are the eigenvalues of* $\mathbf{A}$.

To prove this, we note that R is homogeneous of degree 0 in $\mathbf{x}$. We can write R as the value of $\mathbf{x}^T\mathbf{A}\mathbf{x}$ subject to $\mathbf{x}^T\mathbf{x} = 1$. We can enforce the constraint $\mathbf{x}^T\mathbf{x} = 1$ by introducing a Lagrange multiplier λ, and then investigate the stationary values of

$$R = \mathbf{x}^T\mathbf{A}\mathbf{x} - \lambda\mathbf{x}^T\mathbf{x}.$$

Since

$$\frac{\partial R}{\partial x_i} = 2(a_{1i}x_1 + a_{2i}x_2 + \cdots a_{ni}x_n) - 2\lambda x_i$$

the equations $\partial R/\partial x_i$ for $i = 1, 2, \ldots, n$ may be collected to give

$$\mathbf{A}\mathbf{x} - \lambda\mathbf{x} = \mathbf{0}$$

which states that λ is an eigenvalue of $\mathbf{A}$.

Now return to the need to provide more data for the inverse problem, and suppose that we know the eigenvalues $(\mu_i)_1^{n-1}$ of the matrix $\mathbf{A}$ obtained from $\mathbf{A}$ by deleting row 1 and column 1. The μ_i are the stationary values of

R subject to the constraint $x_1 = 0$. Since $x_1 = 0$ is equivalent to $\mathbf{x}^T\mathbf{e}_1 = 0$, where $\mathbf{e}_1 = \{1, 0, \ldots, 0\}$, we can introduce two Lagrange multipliers λ and 2μ, and consider the stationary values of

$$R = \mathbf{x}^T\mathbf{A}\mathbf{x} - \lambda\mathbf{x}^T\mathbf{x} - 2\mu\mathbf{x}^T\mathbf{e}_1.$$

Now

$$\frac{\partial R}{\partial x_i} = 2\sum_{j=1}^{n} a_{ij}x_j - 2\lambda x_i - 2\mu\delta_{i1} = 0$$

which may be combined to give

$$\mathbf{A}\mathbf{x} - \lambda\mathbf{x} - \mu\mathbf{e}_1 = \mathbf{0}.$$

Since the $(\mathbf{x}_i)_1^n$ span R_n, we may write $\mathbf{x} = \sum_{i=1}^n \alpha_i\mathbf{x}_i$ so that

$$\sum_{i=1}^{n}(\lambda_i - \lambda)\alpha_i\mathbf{x}_i - \mu\mathbf{e}_1 = 0.$$

Using the orthonormality of the $\mathbf{x}_i$, we deduce

$$(\lambda_i - \lambda)\alpha_i = \mu\mathbf{x}_i^T\mathbf{e}_1 = \mu x_{1i}$$

where x_{1i} denotes the first component of x_i. So

$$\alpha_i = \mu x_{1i}/(\lambda_i - \lambda)$$

and the constraint $x_1 = 0$ gives the equation

$$f(\lambda) \equiv \sum_{i=1}^{n}\frac{[x_{1i}]^2}{\lambda_i - \lambda} = 0.$$

This means the eigenvalues $(\mu_i)_1^{n-1}$ are the roots of this equation. If each x_{1i} is non-zero, the graph of $f(\lambda)$ is as shown in Figure 1. The eigenvalues $(\mu_i)_1^{n-1}$ *interlace* the eigenvalues $(\lambda_i)_1^n$:

$$\lambda_1 < \mu_1 < \lambda_2 < \mu_2 < \ldots \mu_{n-1} < \lambda_n. \tag{2}$$

(If, for a certain i, $x_{1i} = 0$, then λ_i itself is an eigenvalue of $\mathbf{A}_1$.) This analysis shows that if we know the $(x_{1i})_1^n$ (and since the values $x_{1i}, x_{2i}, \ldots, x_{ni}$ form a unit vector, $\sum_{i=1}^n [x_{1i}]^2 - 1$) then we can find the $(\mu_i)_1^{n-1}$. But the converse is true: if we know the $(\mu_i)_1^{n-1}$, and they interlace the $(\lambda_i)_1^n$, then we can find the $(x_{1i})_1^n$. For then

$$\sum_{i=1}^{n}\frac{[x_{1i}]^2}{\lambda_i - \lambda} = \frac{\prod_{i=1}^{n-1}(\mu_i - \lambda)}{\prod_{i=1}^{n\prime}(\lambda_i - \lambda)}$$

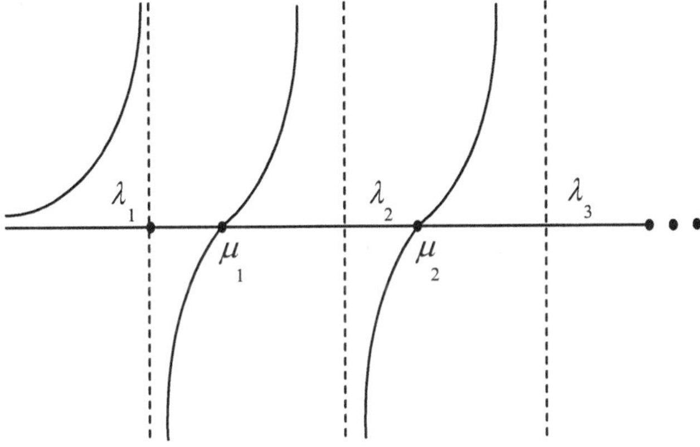

Figure 1. $f(\lambda)$ has roots $\mu_1, \mu_2, \ldots$.

from which we deduce

$$[x_{1i}]^2 = \frac{\prod_{j=1}^{n-1}(\mu_j - \lambda_i)}{\prod_{j=1}^{n\prime}(\lambda_j - \lambda_i)} \tag{3}$$

where $'$ denotes $j \neq i$. We can verify that *if* the μ_j interlace the λ_i, as in (2), *then* $[x_{1i}]^2$ given by (3) is positive.

We conclude that if we know $(\mu_i)_1^{n-1}$ then we know the first entries in the vectors $(\mathbf{x}_i)_1^n$. If we know the $(\lambda_i)_1^n$ and the $(x_{1i})_1^n$, we have $n+n-1 = 2n-1$ pieces of data. (Note that since $\sum_{i=1}^n [x_{1i}]^2 = 1$, the $(x_{1i})_1^n$ provide only $n-1$ pieces of data.)

1.3 Inverse Problems for a Jacobian Matrix

A Jacobian matrix is a matrix in S_n with just three diagonals:

$$\mathbf{J} = \begin{bmatrix} a_1 & b_1 & & & \\ b_1 & a_2 & b_2 & & \\ & \ddots & \ddots & \ddots & \\ & & \ddots & \ddots & b_{n-1} \\ & & & b_{n-1} & a_n \end{bmatrix}.$$

Since changing the sign of one of the b_i has no effect on the eigenvalues, we can assume that $(b_i)_1^{n-1} > 0$ or $(b_i)_1^{n-1} < 0$. If $b_i = 0$ for some $i =$

$1, 2, \ldots, n-1$, then the eigenvalue problem splits into two separate problems, one for $j = 1, \ldots, i$ and one for $j = i+1, \ldots, n$. There is therefore no loss in generality in assuming $(b_i)_1^{n-1} \neq 0$.

If $\mathbf{x}$ is an eigenvector of $\mathbf{J}$, then the first line of $(\mathbf{J} - \lambda\mathbf{I})\mathbf{x} = \mathbf{0}$ is $(a_1 - \lambda)x_1 + b_1 x_2 = 0$. If $x_1 = 0$ then $x_2 = 0$, and then line 2 gives $x_3 = 0$, and so on until $x_n = 0$: $\mathbf{x} = \mathbf{0}$, contradicting the statement that $\mathbf{x}$ is an eigenvector. We deduce that $(x_{1i})_1^n \neq 0$. Moreover, as we showed in Section 1, if we know $(\lambda_i)_1^n$, and $(\mu_i)_1^{n-1}$ and they interlace, then we can find $(x_{1i})_1^n$.

We can assemble the n column-vector equations $(\mathbf{A} - \lambda_i\mathbf{I})\mathbf{x}_i = \mathbf{0}$ into one matrix equation

$$\mathbf{AX} = \mathbf{X}\wedge$$

where

$$\mathbf{X} = \{\mathbf{x}_1, \mathbf{x}_2, \ldots, \mathbf{x}_n\}.$$

We know the first entries in each vector $\mathbf{x}_i$, i.e., the first row of $\mathbf{X}$. If we transpose $\mathbf{X}$ to form $\mathbf{U}$, i.e., $\mathbf{X}^T = \mathbf{U}$, then

$$\mathbf{AU}^T = \mathbf{U}^T\wedge \text{ or, on transposing, } \mathbf{UA} = \wedge\mathbf{U}.$$

Now write $\mathbf{U} = \{\mathbf{u}_1, \mathbf{u}_2, \ldots, \mathbf{u}_n\}$. Since the first column $\mathbf{u}_i$ contains the entries in the first row of $\mathbf{X}$, we know $\mathbf{u}_1$. Write $\mathbf{UA} = \wedge\mathbf{U}$ as

$$\{\mathbf{u}_1, \mathbf{u}_2, \ldots, \mathbf{u}_n\} \begin{bmatrix} a_1 & b_1 & & & \\ b_1 & a_2 & b_2 & & \\ & \ddots & \ddots & \ddots & \\ & & \ddots & \ddots & b_{n-1} \\ & & & b_{n-1} & a_n \end{bmatrix} = \wedge\{\mathbf{u}_1, \mathbf{u}_2, \ldots, \mathbf{u}_n\}. \quad (4)$$

The first column of this equation is

$$a_1\mathbf{u}_1 + b_1\mathbf{u}_2 = \wedge\mathbf{u}_1. \quad (5)$$

Since the columns $\mathbf{u}_1, \mathbf{u}_2, \ldots, \mathbf{u}_n$ are orthonormal, we deduce

$$a_1 = \mathbf{u}_1^T\wedge\mathbf{u}_1.$$

Now write (5) as

$$b_1\mathbf{u}_2 = \wedge\mathbf{u} - a_1\mathbf{u}_1 = \mathbf{z}_2.$$

The vector $\mathbf{z}_2$ is known; $\mathbf{u}_2$ is to be a unit vector, so that

$$b_i^2 = ||\mathbf{z}_2||^2,\ b_1 = ||\mathbf{z}_2||$$

and

$$\mathbf{u}_2 = \mathbf{z}_2/b_1.$$

Now take the second column of (4):

$$b_1\mathbf{u}_1 + a_2\mathbf{u}_2 + b_2\mathbf{u}_3 = \wedge\mathbf{u}_2.$$

Proceeding as before, we find

$$a_2 = \mathbf{u}_2^T\wedge\mathbf{u}_2$$

$$b_2\mathbf{u}_3 = \wedge\mathbf{u}_2 - b_1\mathbf{u}_1 - a_2\mathbf{u}_2 = \mathbf{z}_3$$

from which we deduce

$$b_2 = ||\mathbf{z}_3||, \quad \mathbf{u}_3 = \mathbf{z}_3/b_2.$$

We continue this procedure to find $(a_i)_1^n$ and $(b_i)_1^{n-1}$. This procedure is called the *Lanczos* algorithm. (Note that we have taken the b_i to be positive. If instead the b_i are negative, then $b_1 = -||z_2||$, $b_2 = -||z_3||$, and so on.)

Actually, what we have described is an inverse version of the original Lanczos algorithm. The original algorithm solved the following problem: Given a symmetric matrix $\mathbf{A}$ and a vector $\mathbf{x}_1$ such that $\mathbf{x}_1^T\mathbf{x}_1 = 1$, compute a Jacobi matrix $\mathbf{J}$ and an orthogonal matrix $\mathbf{X} = \{\mathbf{x}_1, \mathbf{x}_2, \ldots, \mathbf{x}_n\}$ such that $\mathbf{A} = \mathbf{XJX}^T$. In our use of the algorithm, we start with $\mathbf{A} = \wedge$.

For variants of the Lanczos algorithm, and alternative ways to deduce the entries in a Jacobi matrix from eigenvalue data, see Gladwell (2004), especially Sections 4.2 and 4.3. See Golub (1973) for an early paper on this algorithm.

It may be shown that the inverse problem of constructing $\mathbf{J}$ from data $(\lambda_i)_1^n$ and $(x_{1i})_1^n$ is well conditioned, but some ways of carrying out the computation are better conditioned than others. This is discussed at length in the works just cited. See also Boley and Golub (1984) and Boley and Golub (1987).

The inverse eigenvalue problem for a Jacobi matrix can be linked to a mechanical inverse problem: find the masses $(m_i)_1^n$ and spring stiffnesses $(k_i)_1^n$ in the system shown in Figure 2, so that the system has natural frequencies $(\omega_i)_1^n$ and, when the right-hand end is fixed, has natural frequencies $(\omega_i^*)_1^{n-1}$, where $(\omega_i)_1^n$ and $(\omega_i^*)_1^{n-1}$ interlace. (Notice that for convenience, we have turned the system round so that the right-hand end is first free and then fixed: when considering A, we considered A and A_1, the matrix obtained by deleting row 1 and column 1.) The equation for the natural frequencies $(\omega_i)_1^n$ is

$$(\mathbf{K} - \lambda\mathbf{M})\mathbf{u} = \mathbf{0}, \quad \lambda = \omega^2,$$

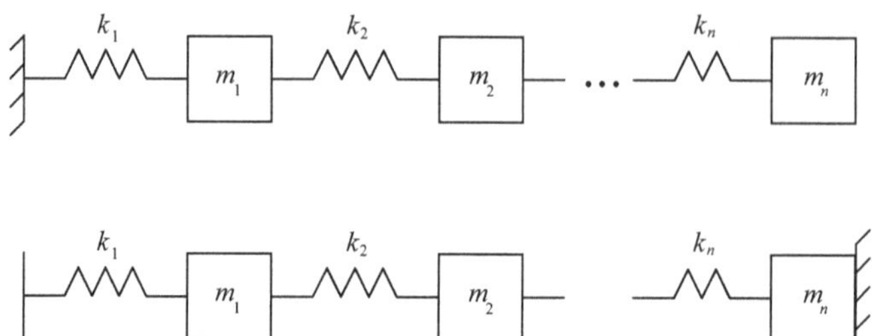

Figure 2. The mass-spring system has its right-hand end free (above) or fixed (below).

where

$$\mathbf{K} = \begin{bmatrix} k_1+k_2 & -k_2 & & & \\ -k_2 & k_2+k_3 & \ddots & & \\ & \ddots & \ddots & \ddots & \\ & & \ddots & \ddots & -k_n \\ & & & -k_n & k_n \end{bmatrix}, \quad \mathbf{M} = diag(m_1, m_2, \ldots, m_n).$$

To reduce the problem to standard form, write

$$\mathbf{M} = \mathbf{D}^2, \quad \mathbf{D}^{-1}\mathbf{K}\mathbf{D}^{-1} = \mathbf{A}, \quad \mathbf{D}\mathbf{u} = \mathbf{x} \tag{6}$$

so that

$$(\mathbf{A} - \lambda\mathbf{I})\mathbf{x} = \mathbf{0}.$$

Having found the Jacobi matrix $\mathbf{A}$ by using the Lanczos algorithm, we have to isolate $\mathbf{K}$ and $\mathbf{M}$ from $\mathbf{A}$. Equation (6) shows that $\mathbf{K} = \mathbf{DAD}$. By inspection, $\mathbf{K}$ has the property

$$\mathbf{K}\{1, 1, \ldots, 1\} = k_1\mathbf{e}_1$$

where $\mathbf{e}_1 = \{1, 0, \ldots, 0\}$. Therefore,

$$\mathbf{DAD}\{1, 1, \ldots, 1\} = k_1\mathbf{e}_1$$

and hence

$$\mathbf{A}\{d_1, d_2, \ldots, d_n\} = \mathbf{D}^{-1}k_1\mathbf{e}_1 = d_1^{-1}k_1\mathbf{e}_1. \tag{7}$$

We now proceed as follows. Take the unique reconstructed Jacobi matrix $\mathbf{A}$, (we choose the b_i to be *negative*, just as the off-diagonal entries of $\mathbf{K}$ are negative) and solve

$$\mathbf{A}\{x_1, x_2, \ldots, x_n\} = \mathbf{e}_1.$$

We can show that the solution, $\mathbf{x}$, is *strictly positive*: $x_1, x_2, \ldots, x_n$ are all positive. The solution of equation (6) is $\mathbf{d} = c\mathbf{x}$ for some as yet unknown c. The total mass of the system is

$$m = \sum_{i=1}^{n} m_i = \sum_{i=1}^{n} d_i^2 = ||\mathbf{d}||^2 = c^2||\mathbf{x}||^2.$$

Thus, choosing the total mass, and computing $||\mathbf{x}||^2$, we can find c, and hence $\mathbf{d}$, and hence the masses $m_i = d_i^2$. Now we compute $\mathbf{K} = \mathbf{DAD}$ which yields the stiffness k_i.

2 Lecture 2. Applications and Extensions

2.1 A Minimal Mass Problem

In Lecture 1, we showed that a Jacobi matrix $\mathbf{J}$ could be reconstructed uniquely from $(2n-1)$ pieces of data: the eigenvalues of $\mathbf{J}$, and of $\mathbf{J}_1$, the matrix obtained by deleting row 1 and column 1 of $\mathbf{J}$. The eigenvalues must interlace in the sense

$$\lambda_1 < \mu_1 < \lambda_2 < \ldots \mu_{n-1} < \lambda_n. \tag{8}$$

An equivalent set of data is $(\lambda_i)_1^n$ and $(x_{1i})_1^n$. Here x_{1i} is the first component of the ith eigenvector $\mathbf{x}_i$; the x_{1i} are normalized so that

$$\sum_{i=1}^{n} [x_{1i}]^2 = 1.$$

In Lecture 1, we showed how this inversion procedure could be used to construct a spring mass system such as that shown in Figure 2. Gladwell (2006) used this analysis to solve a minimal mass problem: of all possible spring-mass systems as shown in Figure 2(a), which have the given spectrum $(\lambda_i)_1^n$, find the one that minimizes the mass $m = \sum_{i=1}^{n} m_i$ for given stiffness k given by

$$\frac{1}{k} = \sum_{i=1}^{n} \frac{1}{k_i}.$$

Since the system can be reconstructed from $(\lambda_i)_1^n$ and the *last* entries u_{ni} in the eigenvectors, the problem reduces to finding those u_{ni} that minimize m/k.

A somewhat intricate analysis shows that

$$\frac{m}{k} = \left(\prod_{i=1}^{n} \lambda_i\right)^2 \cdot \sum_{i=1}^{n} \{\lambda_i^2 D_i^2 (u_{ni})^2\}^{-1} \cdot \sum_{i=1}^{n} \lambda_i^{-1} (u_{ni})^2, \tag{9}$$

where

$$D_i = \prod_{j=1}^{n} {}'(\lambda_i - \lambda_j), \text{ and } ' \text{ denotes } i \neq j.$$

To find the global minimum of m/k, we note that (9) is a homogeneous function of degree zero in the u_{ni}. To find its minimum, we use the Cauchy-Schwarz inequality $||\mathbf{x}||^2||\mathbf{y}||^2 \geq |\mathbf{x} \cdot \mathbf{y}|^2$ to deduce that the global minimum occurs when

$$\lambda_i^{-1}[u_{ni}]^2 = c^2(\lambda_i^2 D_i^2 [u_{ni}]^2)^{-1},$$

for some constant c. Thus

$$(u_{ni})^4 = c^2 \lambda_i^{-1} D_i^{-2}, \quad [u_{ni}]^2 = c\lambda_i^{-1/2} D_i^{-1}, \tag{10}$$

and the minimum is

$$\left(\frac{m}{k}\right)_{\min} = \left(\prod_{i=1}^{n} \lambda_i\right)^2 \left(\sum_{i=1}^{n} (\lambda_i^{3/2} D_i)^{-1}\right)^2.$$

Since the minimizing values of the u_{ni} are known in terms of the eigenvalues data, (see (9)), the values of the individual masses m_i and stiffness k_i may be found as in Section 1.3.

This minimal mass problem is interesting because it shows that it is possible to set up an inverse eigenvalue problem for a spring-mass system for which the given data is *one* spectrum not two.

Note that it is *not* possible to construct a Jacobi matrix from its eigenvalues $(\lambda_i)_1^n$ and one *eigenvector* $\mathbf{x}_i$ using a variant of the Lanczos algorithm. It is possible to construct a spring mass system which has a given mode shape for ith mode, but if this reconstruction is to be realistic, that is, to have positive masses and stiffnesses, then the ith mode shape $\mathbf{u}_i$ must satisfy certain strict conditions on the sign properties of the sequence $u_{1i}, u_{2i}, \ldots, u_{ni}$. One condition is that $u_{1i} \neq 0$, $u_{ni} \neq 0$ as we found in Section 1.3. Also, if one $u_{ji} = 0$ then the neighbouring entries $u_{j-1,i}$ and $u_{j+1,i}$ must be strictly non-zero and have opposite signs. When these conditions are fulfilled then we can give a precise definition for the number of sign changes in the sequence $u_{1i}, u_{2i}, \ldots, u_{ni}$; the ith mode must have exactly $i-1$ sign changes. Thus, the first mode is positive, the second has one sign change, and so on. Not only must $\mathbf{u}_i$ have this property, but also the $\mathbf{w}_i$ where $\mathbf{w} = \mathbf{E}^T\mathbf{u}$, and

$$\mathbf{E} = \begin{bmatrix} 1 & i & & & \\ & 1 & i & & \\ & & \ddots & \ddots & \\ & & & \ddots & i \\ & & & & 1 \end{bmatrix} \tag{11}$$

where $i = -1$. Thus, $w_{1i} = u_{1i}, w_{2i} = u_{2i} - u_{1i}, \ldots, w_{ni} = u_{ni} - u_{n-1,i}$. It can be shown (Gladwell (2004), p. 207) that if $u_{1j}w_{1j} > 0$, and the sequences $(w_{ij})_{i=1}^n$ and $(u_{ij})_{i=1}^n$ have exactly $j-1$ sign changes, then there is a spring mass system of the form in Figure 2(a) that has $(u_{ij})_{i=1}^n$ as its jth mode. It can be shown that there is a family of such systems.

If two mode shapes are known, then providing that much more stringent conditions are satisfied, again it is possible to construct a spring-mass system which has those modes as eigenmodes; under certain conditions, this system is uniquely determined. See Vijay (1972) and Gladwell (2004).

2.2 Another Spring-Mass Inverse Problem

The inverse problem we have studied involves finding the stiffnesses $(k_i)_1^n$ and the masses $(m_i)_1^n$ from two spectra $(\omega_i)_1^n$ and $(\omega_i^*)_1^n$. We note that multiplying all the stiffnesses and masses by a constant c does not alter the spectra, there are only $2n-1$ ratios to be found; we can fix one of the quantities $(k_i, m_i)_1^n$, or fix, say, the total mass $m = \sum m_i$. It is tempting to suppose that an easier inverse problem would be to fix the masses $(m_i)_1^n$, and find the n stiffnesses $(k_i)_1^n$ from one spectrum $(\omega_i)_1^n$. However, even for $n = 2$, it is not certain that there is a system with given $m_1, m_2, \lambda_1, \lambda_2$, (where $\lambda_i = \omega_i^2$). The eigenvalues are the roots of

$$\begin{vmatrix} k_1 + k_2 - \lambda m_1 & -k_2 \\ -k_2 & k_2 - \lambda m_2 \end{vmatrix} = 0.$$

If the eigenvalues λ_1, λ_2 and the masses m_1, m_2 are known, then $k_1 = (m_1 m_2 \lambda_1 \lambda_2)/k_2$, and k_2 is a root of the quadratic equation

$$\left(\frac{m_1 + m_2}{m_1 m_2}\right) k_2^2 - (\lambda_1 + \lambda_2)k_2 + m_2\lambda_1\lambda_2 = 0,$$

which has real positive roots if and only if

$$(\lambda_1 - \lambda_2)^2 > 4\frac{m_2}{m_1}\lambda_1\lambda_2.$$

This essentially states that the eigenvalues should not be too close to each other. For $n = 3$ the analysis becomes formidable, and essentially intractable: it is necessary to find the conditions on the data which will ensure that a certain sixth order polynomial equation has a real positive root. This problem is very similar to the so-called *inverse additive eigenvalue problem*: Given $\mathbf{A} \in S_n$, find a diagonal matrix $\mathbf{D}$ such that $\mathbf{A} + \mathbf{D}$ has eigenvalues $(\lambda_i)_1^n$; see Chu and Golub (2005), Friedland (1977). It is somewhat curious that the inverse problem in which the $2n$ quantities $(k_i, m_i)_1^n$ are to be found is much simpler than that in which only $(k_i)_1^n$ are to be found. In a way, this should not be surprising because the quantity in equation (2) is (a multiple of) the response function of the system, the data $(\lambda_i)_1^n$ and $(\mu_i)_1^n$ are the poles and zeros of this response function and it is known that the poles and zeros of the response function must interlace. Essentially, the system is reconstructed from its response function.

2.3 Inverse Problems for a Pentadiagonal Matrix

In Section 1.3, we showed that we could construct a Jacobian matrix $\mathbf{A}$ (a tridiagonal matrix) from the spectrum $(\lambda_i)_1^n$, making up $\wedge$, and the first (or last) components of the normalized eigenvectors of $\mathbf{A}$. We can easily extend the Lanczos algorithm to the reconstruction of a pentadiagonal matrix $\mathbf{A}$ from its eigenvalues, making up $\wedge$, and the first *two* components of the normalized eigenvectors of $\mathbf{A}$. Now equation (4) is replaced by

$$[\mathbf{u}_1, \mathbf{u}_2, \ldots, \mathbf{u}_n] \begin{bmatrix} a_i & b_1 & c_1 & & & & \\ b_1 & a_2 & b_2 & c_2 & & & \\ c_1 & b_2 & a_3 & b_3 & c_3 & & \\ & & & \ddots & \ddots & & \\ & & & & \ddots & \ddots & c_{n-2} \\ & & & & & \ddots & b_{n-1} \\ & & & & & & a_n \end{bmatrix} = \wedge[\mathbf{u}_1, \mathbf{u}_2, \ldots, \mathbf{u}_n].$$

The first column of this equation is

$$a_1\mathbf{u}_1 + b_1\mathbf{u}_2 + c_1\mathbf{u}_3 = \wedge\mathbf{u}_1.$$

If $\mathbf{u}_1, \mathbf{u}_2$ are known (and they must satisfy $\mathbf{u}_i^T\mathbf{u}_j = \delta_{ij}$, for $i, j = 1, 2$) then

$$a_1 = \mathbf{u}_1^T\wedge\mathbf{u}_1, \quad b_1 = \mathbf{u}_2^T\wedge\mathbf{u}_1,$$

and

$$c_1\mathbf{u}_3 = \Lambda\mathbf{u}_1 - a_1\mathbf{u}_1 - b_1\mathbf{u}_2 = \mathbf{z}_3,$$

and $||\mathbf{u}_3|| = 1$ gives

$$c_1 = ||\mathbf{z}_3||, \quad \mathbf{u}_3 = \mathbf{z}_3/c_1, \quad \text{and so on.}$$

Formally, this analysis appears to be correct, but in practice, the matrix that is being reconstructed has a particular form, and in particular, has sign properties.

A typical case in which this pentadiagonal inverse problem arises is the inverse problem of a beam modelled as point masses attached by torsional springs to rigid rods. Now the stiffness matrix $\mathbf{K}$ has the form

$$\mathbf{K} = \mathbf{E}\mathbf{L}^{-1}\mathbf{E}\hat{\mathbf{K}}\mathbf{E}^T\mathbf{L}^{-1}\mathbf{E}^T,$$

where

$$\hat{\mathbf{K}} = diag(k_1, k_2, \ldots, k_n), \quad \mathbf{L} = diag(\ell_1, \ell_2, \ldots, \ell_n)$$

and $\mathbf{E}$ is given by (11). The eigenvalue problem is $(\mathbf{K} - \lambda\mathbf{M})\mathbf{x} = \mathbf{0}$, which reduces to $(\mathbf{A} - \lambda\mathbf{I})\mathbf{y} = \mathbf{0}$ where $\mathbf{A} = \mathbf{D}^{-1}\mathbf{K}\mathbf{D}^{-1}$, and $\mathbf{M} = \mathbf{D}^2$. The data must be chosen so that all the quantities $(k_i, m_i, \ell_i)_1^n$ are *positive*. The necessary conditions on the data, essentially making up $\mathbf{u}_1$ and $\mathbf{u}_2$, were found by Gladwell (1984), Gladwell (2004). Finding the conditions on the data is actually much harder than the formal reconstruction of the matrix $\mathbf{A}$, and hence $(k_i, m_i, \ell_i)_1^n$.

2.4 Periodic Jacobi Matrices

A Jacobian matrix is a tridiagonal matrix of the form (4). A *periodic* Jacobi matrix has the form

$$\mathbf{J}_{per} = \begin{bmatrix} a_1 & b_1 & & & b_n \\ b_1 & a_2 & b_2 & & \\ & \ddots & \ddots & \ddots & \\ & & \ddots & \ddots & b_{n-1} \\ b_n & & & b_{n-1} & a_n \end{bmatrix}.$$

It is called periodic because it is as if it were inscribed on a cylinder, and b_n appears to the left of a_1 and to the right of a_n. The graph underlying $\mathbf{J}_{per}$ is a ring on n vertices.

Boley and Golub (1987) showed that $\mathbf{J}_{per}$ could be reconstructed from $(\lambda_i)_1^n$, the eigenvalues of $\mathbf{J}_{per}$; $(\mu_i)_1^{n-1}$, the eigenvalues of $(\mathbf{J}_{per})_1$, and one extra piece of data, $\beta = b_1 b_2 \dots b_n$. Now, however, there are conditions on the data which will ensure that there is a real solution; these were found by Xu (1998); see also Boley and Golub (1984) and Andrea and Berry (1992).

2.5 Graph Theory

The proper study of matrix form, i.e., the pattern of zero and non-zero entries in a matrix, requires a knowledge of *graph* theory. A *graph* $\mathcal{G}$ is a set of *vertices*, P_i, connected by *edges*, (P_i, P_j). The set of vertices is called the *vertex* set, and is denoted by $\mathcal{V}$; the set of edges is called the *edge* set $\mathcal{E}$. A graph is said to be *simple* if there is at most one edge connecting any two vertices. The graph is *undirected* if there is no preferred direction associated with an edge. With any symmetric matrix $\mathbf{A}$ we may associate an undirected graph; the rule is

$$\text{if } i \neq j \text{ then } (P_i, P_j) \in \mathcal{E} \text{ iff } a_{ij} \neq 0. \tag{12}$$

A matrix $\mathbf{A} \in S_n$ is said to be *irreducible* if it *cannot* be reduced to the form

$$\mathbf{A} = \left[\begin{array}{c|c} \mathbf{B} & \\ \hline & \mathbf{C} \end{array}\right]$$

by any rearrangement of rows and columns.

A graph $\mathcal{G}$ is said to be *connected* if there is a chain consisting of a sequence of edges connecting any one vertex to any other vertex. It can be shown that $\mathbf{A} \in S_n$ is *irreducible* if and only if its underlying graph is *connected*.

A *tree* is a connected graph with no cycles. The simplest tree is a *path*, a set of vertices $(P_i)_1^n$ with $n-1$ edges $(P_1, P_2), (P_2, P_3), \dots, (P_{n-1}, P_n)$. A Jacobi matrix is associated with a path, while that associated with a periodic Jacobi matrix is a cycle with edges $(P_1, P_2), (P_2, P_3), \dots, (P_{n-1}, P_n), (P_n, P_1)$.

Duarte (1989), in an important paper, solved the inverse problem for a tree, i.e., for a matrix $\mathbf{A} \in S_n$ whose underlying matrix is a tree. He showed that, given two interlacing spectra $(\lambda_i)_1^n$ and $(\mu_i)_1^{n-1}$, it is possible to find a matrix $\mathbf{A}$ with underlying graph a tree, such that $(\lambda_i)_1^n$ and $(\mu_i)_1^{n-1}$ are the spectra of $\mathbf{A}$ and $\mathbf{A}_1$ respectively. For an example, consider the tree shown in Figure 3, for which $\mathbf{A}$ has the form

$$\mathbf{A} = \begin{bmatrix} x & x & & & & x & & & \\ x & x & x & x & x & & & & \\ & x & x & & & & & & \\ & x & & x & x & & & & \\ & x & & & x & & & & \\ x & & & & & x & x & x & \\ & & & & & x & x & & \\ & & & & & x & & x & x \\ & & & & & & & x & x \end{bmatrix}$$

Given $(\lambda_i)_1^n$ and $(\mu_i)_1^n$, we find a_{11}, a_{12}, a_{16} and we find sufficient new

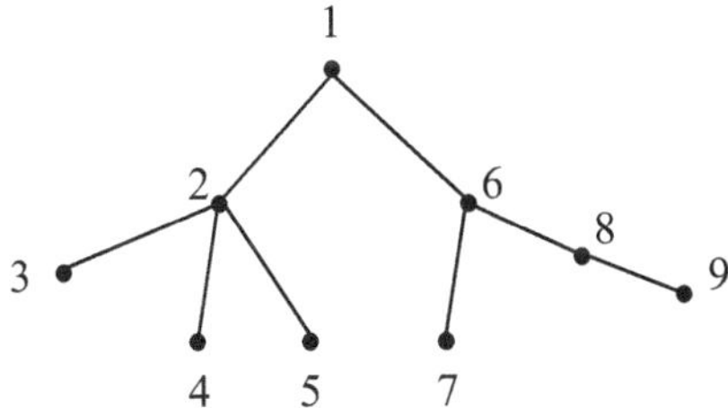

Figure 3. The graph underlying $\mathbf{A}$ is a tree.

data to give $a_{22}, a_{23}, a_{24}, a_{25}$ on the one hand and a_{66}, a_{67}, a_{68} on the other. Finally, we find a_{88}, a_{89}. Further discussion of, and references to, eigenvalue problems related to trees may be found in Nylen and Uhlig (1994).

The defining characteristic of a path is that the graph has an *end*; one can start at this end and proceed to the other end. For a tree, there is also an end, a root. One can start at this end and proceed to a vertex at which the tree splits into two or more subtrees; again one can proceed along each of these subtrees, and so finally reach each vertex.

Unfortunately, the stiffness matrices obtained through a finite element model of a two- or three- dimensional solid structure are in general not trees - they are based on triangulation or some other similar process. There is no 'end' of the underlying graph from which one can start.

The simplest example of a matrix whose underlying graph is *not* a tree is a periodic Jacobi matrix. Just as the inverse problem for a periodic Jacobi matrix requires that rather curious extra piece of data, β, so the inverse problem for a matrix with underlying triangulated graph requires more data than just two spectra. Such inverse problems are currently still open.

3 Lecture 3. Isospectral Systems

3.1 Reversing Factors

Two vibrating systems that have the same natural frequencies are said to be *isospectral*. Similarly, two matrices $\mathbf{A}, \mathbf{B} \in S_n$ with the same eigenvalues are said to be isospectral.

In Lecture 1, we showed that if $\mathbf{A} \in S_n$ has eigenvalues $(\lambda_i)_1^n$ making up $\wedge = diag(\lambda_1, \lambda_2, \ldots, \lambda_n)$, and normalized eigenvectors $\mathbf{x}_i$ making up $\mathbf{X} = \{\mathbf{x}_1, \mathbf{x}_2, \ldots, \mathbf{x}_n\}$, then

$$\mathbf{AX} = \mathbf{X}\wedge, \quad \mathbf{XX}^T = \mathbf{X}^T\mathbf{X} = \mathbf{I},$$

so that

$$\mathbf{A} = \mathbf{X}\wedge\mathbf{X}^T.$$

If $\mathbf{B} \in S_n$ also has eigenvalues $(\lambda_i)_1^n$, then

$$\begin{aligned} \mathbf{B} &= \mathbf{Y}\wedge\mathbf{Y}^T, \\ &= \mathbf{YX}^T(\mathbf{X}\wedge\mathbf{X}^T)\mathbf{XY}^T = \mathbf{QAQ}^T, \end{aligned}$$

where $\mathbf{Q} = \mathbf{YX}^T$ is an orthogonal matrix ($\mathbf{QQ}^T = \mathbf{YX}^T(\mathbf{XY}^T) = \mathbf{YY}^T = \mathbf{I}$).

Conversely, if $\mathbf{B} = \mathbf{QAQ}^T$ and $\mathbf{AX} = \mathbf{X}\wedge$, then

$$\mathbf{B}(\mathbf{QX}) = \mathbf{QAQ}^T(\mathbf{QX}) = \mathbf{QAX} = (\mathbf{QX})\wedge,$$

so that $\mathbf{B}$ has the same eigenvalues as $\mathbf{A}$.

This shows that all the matrices $\mathbf{B} \in S_n$ that are isospectral to $\mathbf{A}$ are given by $\mathbf{B} = \mathbf{QAQ}^T$ for some orthogonal matrix $\mathbf{Q}$.

However, if $\mathbf{A}$ has some special form, e.g., tridiagonal, pentadiagonal, then $\mathbf{B} = \mathbf{QAQ}^T$ will share this form only for special orthogonal matrices $\mathbf{Q}$.

To obtain a class of matrices $\mathbf{B} \in S_n$ that are isospectral to $\mathbf{A} \in S_n$, we recall a simple but far-reaching result.

If $\mathbf{A}, \mathbf{B} \in M_n$, *then* $\mathbf{AB}$ *and* $\mathbf{BA}$ *have the same eigenvalues, except perhaps for zero.* To prove this, suppose $\lambda \neq 0$ is an eigenvalue of $\mathbf{AB}$ so that, for some $\mathbf{x} \neq \mathbf{0}$, $\mathbf{ABx} = \lambda\mathbf{x}$. Since $\lambda \neq 0$ and $\mathbf{x} \neq \mathbf{0}$, we have $\mathbf{Bx} \neq \mathbf{0}$, and $\mathbf{B}(\mathbf{ABx}) = \lambda\mathbf{Bx}$, i.e., $\mathbf{BA}(\mathbf{Bx}) = \lambda\mathbf{Bx}$. λ is an eigenvalue of $\mathbf{BA}$ with eigenvector $\mathbf{Bx}$. Now reverse the roles of $\mathbf{A}$ and $\mathbf{B}$ to complete the proof.

To use this result to obtain a system that is isospectral to the system show in Figure 2(a), write $\mathbf{K}$ and $\mathbf{M}$ of equation (7) in the form

$$\mathbf{K} = \mathbf{E}\hat{\mathbf{K}}\mathbf{E}^T, \quad \hat{\mathbf{K}} = \mathbf{F}^2, \quad \mathbf{M} = \mathbf{D}^2,$$

where

$$\mathbf{F} = diag(f_1, f_2, \ldots, f_n), \quad \mathbf{D} = diag(d_1, d_2, \ldots, d_n).$$

Thus $(\mathbf{K} - \lambda\mathbf{M})\mathbf{u} = \mathbf{0}$ is reduced to $(\mathbf{A} - \lambda\mathbf{I})\mathbf{x} = \mathbf{0}$, where

$$\mathbf{A} = \mathbf{D}^{-1}\mathbf{E}\mathbf{F}^2\mathbf{E}^T\mathbf{D}^{-1} = (\mathbf{D}^{-1}\mathbf{E}\mathbf{F})(\mathbf{F}\mathbf{E}^T\mathbf{D}^{-1}).$$

Now reverse the factors in $\mathbf{A}$ to form $\mathbf{B}$:

$$\mathbf{B} = (\mathbf{F}\mathbf{E}^T\mathbf{D}^{-1})(\mathbf{D}^{-1}\mathbf{E}\mathbf{F}).$$

Thus matrix $\mathbf{B}$ is a Jacobi matrix; $\mathbf{D}^{-1}$ and $\mathbf{F}$ have been interchanged, as have $\mathbf{E}$ and $\mathbf{E}^T$. It may be verified that the isospectral system related to $\mathbf{B}$ is a spring mass system free at the left-hand end and fixed at the right. This reversing of factors may be used in numerous contexts.

3.2 The QR Algorithm

If $\mathbf{A} \in M_n$ is non-singular, then it can be factorized in the form

$$\mathbf{A} = \mathbf{Q}\mathbf{R},$$

where $\mathbf{Q}$ is orthogonal, and $\mathbf{R}$ is upper triangular with positive diagonal. $\mathbf{QR}$ factorization is equivalent to applying Gram-Schmidt orthogonalization to the columns of $\mathbf{A}$. If $\mathbf{A} = \mathbf{QR}$, then $\mathbf{B} = \mathbf{RQ} = (\mathbf{Q}^T\mathbf{Q})(\mathbf{RQ}) = \mathbf{Q}^T\mathbf{AQ}$. Thus $\mathbf{A}$ and $\mathbf{B}$ are isospectral, and if $\mathbf{A} \in S_n$, so is $\mathbf{B}$. What is more important, however, is that $\mathbf{A}$ and $\mathbf{B}$ often have the same form. To see this, we note that $\mathbf{A}$ and $\mathbf{B}$ are linked by

$$\mathbf{RA} = \mathbf{BR}. \tag{13}$$

Suppose $\mathbf{A}$ is a Jacobi matrix, as in (4), then

$$\begin{bmatrix} r_{11} & r_{12} & r_{13} & \cdots & r_{1n} \\ & r_{22} & r_{23} & \cdots & r_{2n} \\ & & r_{33} & \cdots & \\ & & & & \\ & & & & r_{nn} \end{bmatrix} \begin{bmatrix} a_1 & b_1 & & & \\ b_1 & a_2 & b_2 & & \\ & \ddots & \ddots & \ddots & \\ & & \ddots & \ddots & b_{n-1} \\ & & & b_{n-1} & a_n \end{bmatrix} =$$

$$\begin{bmatrix} b_{11} & b_{12} & \dots & \dots & b_{1n} \\ b_{21} & b_{22} & \dots & \dots & b_{2n} \\ & & & & \\ & & & & \\ & & & & b_{nn} \end{bmatrix} \begin{bmatrix} r_{11} & r_{12} & \dots & \dots & r_{1n} \\ & r_{22} & \dots & \dots & r_{2n} \\ & & \ddots & & \\ & & & \ddots & \\ & & & & r_{nn} \end{bmatrix}.$$

Column 1 row 2 gives $r_{22}b_1 = r_{11}b_{21}$; since r_{11}, r_{22} are positive, b_{21} has the same *sign* as b_1. If $j \geq 3$ then column 1, row j gives

$$0 = b_{j1}r_{11},$$

which implies $b_{j1} = 0$ for $j \geq 3$, and since $\mathbf{B}$ is symmetric, $b_{1j} = 0$ for $j \geq 3$. This argument may be continued to show that $\mathbf{B}$ is a Jacobi matrix.

QR factorization and reversal may be given more power by introducing a shift, μ. Suppose μ is *not* an eigenvalue of $\mathbf{A} \in S_n$, and that

$$\mathbf{A} - \mu\mathbf{I} = \mathbf{QR}, \tag{14}$$

and

$$\mathbf{B} - \mu\mathbf{I} = \mathbf{RQ}, \tag{15}$$

then

$$\begin{aligned} \mathbf{B} &= \mu\mathbf{I} + \mathbf{RQ} = \mathbf{Q}^T\mathbf{Q}(\mu\mathbf{I} + \mathbf{RQ}), \\ &= \mathbf{Q}^T(\mu\mathbf{I} + \mathbf{QR})\mathbf{Q} = \mathbf{Q}^T\mathbf{AQ}, \end{aligned}$$

and again

$$\mathbf{RA} = \mu\mathbf{R} + (\mathbf{B} - \mu\mathbf{I})\mathbf{R} = \mathbf{BR}. \tag{16}$$

We can use this result to obtain a new way of displaying all the Jacobi matrices isospectral to a given one. First, we recall that we can reconstruct a Jacobi matrix from its eigenvalues and the first (or last) components x_{1i} (or x_{ni}) of the normalized eigenvectors $\mathbf{x}_i$. Since $\sum[x_{1i}]^2 = 1$, we may picture a set of values $(x_{1i})_1^n$ as a point on the unit sphere in R_n. Each point gives a Jacobi matrix.

Now consider the other way. We can consider the two step process (14), (15) as defining an operator $\mathcal{G}_\mu$ which takes $\mathbf{A}$ into $\mathbf{B}$: $\mathbf{B} = \mathcal{G}_\mu\mathbf{A}$. Suppose that $\mathbf{A}$ is a Jacobi matrix with positive off diagonal, then $\mathcal{G}_\mu$ is commutative when applied to $\mathbf{A}$, i.e.,

$$\mathcal{G}_{\mu_1}\mathcal{G}_{\mu_2}\mathbf{A} = \mathcal{G}_{\mu_2}\mathcal{G}_{\mu_1}\mathbf{A}. \tag{17}$$

To prove this, suppose that $\mathbf{u}$ is a normalized eigenvector of $\mathbf{A}$:

$$\mathbf{Au} = \lambda\mathbf{u},$$

then

$$\mathbf{B}\mathbf{Q}^T\mathbf{u} = (\mathbf{Q}^T\mathbf{A}\mathbf{Q})\mathbf{Q}^T\mathbf{u} = \mathbf{Q}^T\mathbf{A}\mathbf{u} = \lambda\mathbf{Q}^T\mathbf{u},$$

so that $\mathbf{u}^* = \mathbf{Q}^T\mathbf{u}$ is a normalized eigenvector of $\mathbf{B}$. We may express this eigenvector in another way. Since

$$\mathbf{A}\mathbf{u} = (\mathbf{Q}\mathbf{R} + \mu\mathbf{I})\mathbf{u} = \lambda\mathbf{u},$$

we have

$$\mathbf{Q}\mathbf{R}\mathbf{u} = (\lambda - \mu)\mathbf{u},$$

or

$$\mathbf{u}^* = \mathbf{Q}^T\mathbf{u} = \frac{\mathbf{R}\mathbf{u}}{\lambda - \mu}.$$

This equation shows that the last component of $\mathbf{u}^*$ is

$$\mathbf{u}_n^* = \frac{r_{nn}u_n}{\lambda - \mu},$$

and the last component of the eigenvector $\mathbf{u}_i^*$ may be taken to be

$$\mathbf{u}_{ni}^* = \frac{r_{nn}(\mu)u_{ni}}{|\lambda_i - \mu|}. \tag{18}$$

This shows that, under the operation $\mathcal{G}_\mu$, the last components of the eigenvectors are simply multiplied by two terms, one, $r_{nn}(\mu)$, independent of i, and the other $|\lambda_i - \mu|^{-1}$. This means that the last components of the normalized eigenvectors of either of the two matrices in (16) will be proportional to

$$\frac{u_{ni}}{|\lambda_i - \mu_1||\lambda_i - \mu_2|}.$$

Since they are proportional, and the sum of squares of each set is unity, the two sets must be the same. But a Jacobi matrix is uniquely determined by its eigenvalues and the last components of its normalized eigenvectors. Therefore, (17) holds, and $\mathcal{G}_\mu$ is commutative.

We can now prove that if $\mathbf{A}, \mathbf{B}$ are two isospectral Jacobi matrices with positive off-diagonal terms, then we can find a unique set $(\mu_i)_1^{n-1}$ such that

$$\mu_1 < \mu_2 < \ldots < \mu_{n-1},$$

and

$$\mathcal{G}_{\mu_1}\mathcal{G}_{\mu_2}\ldots\mathcal{G}_{\mu_{n-1}}\mathbf{A} = \mathbf{B}.$$

It is sufficient to show that we can pass from one set of last components $(u_{ni})_1^n$ to any other set $(v_{ni})_1^n$ in $(n-1)$ $\mathcal{G}_\mu$ operations. But equation (16) shows that this is equivalent to choosing $(\mu_j)_1^{n-1}$ such that

$$\prod_{j=1}^{n-1} \frac{u_{ni}}{|\lambda_i - \mu_j|} \quad \alpha v_{ni}, \quad i = 1, 2, \ldots, n.$$

This is equivalent to choosing the polynomial

$$P(\lambda) = K \prod_{j=1}^{n-1} (\lambda - \mu_j),$$

such that

$$|P(\lambda_i)| = u_{ni}/v_{ni} \quad i = 1, 2, \ldots, n.$$

If we choose the μ_j so that

$$\lambda_1 < \mu_1 < \lambda_2 < \ldots < \mu_{n-1} < \lambda_n, \tag{19}$$

then

$$P(\lambda_i) = (-)^{n-i} u_{ni}/v_{ni} \quad i = 1, 2, \ldots, n.$$

But there is a unique such polynomial $P(\lambda)$ of degree $n-1$ having values of opposite signs at n points λ_i, and it will have roots μ_i satisfying (17).

This means that, given $\mathbf{A}, \mathbf{B}$, we may pass from $\mathbf{A}$ to $\mathbf{B}$ in $n-1$ steps $\mathcal{G}_{\mu_j}$, and the order in which we take these steps is immaterial.

We used the equation $\mathbf{RA} = \mathbf{BR}$ to show that if $\mathbf{A}$ is a Jacobi matrix, then so is $\mathbf{B}$. There is a more general result relating to a so-called *staircase* matrix. A symmetric staircase matrix has its non-zero entries clustered around the diagonal as in a staircase, as shown in Figure 4. It is easily shown that if $\mathbf{A} \in S_n$ is a staircase matrix, and $\mathbf{B} = \mathcal{G}_\mu \mathbf{A}$, then $\mathbf{B}$ is a staircase matrix with the same pattern. Arbenz and Golub (1995) showed that staircase patterns are effectively the only ones invariant under the symmetric (unshifted) QR algorithm.

3.3 Positivity

Not only do the matrices appearing in physical problems have a particular pattern of *zero* and *non-zero* entries, but they also have particular patterns of *positive* and *negative* entries. We limit our discussion to symmetric matrices $\mathbf{A} \in S_n$.

The definition *positive definite* is well-known: $\mathbf{A} \in S_n$ is said to be positive definite PD (positive semi-definite PSD) if $\mathbf{x}^T\mathbf{A}\mathbf{x} > 0 (\geq 0)$ for all

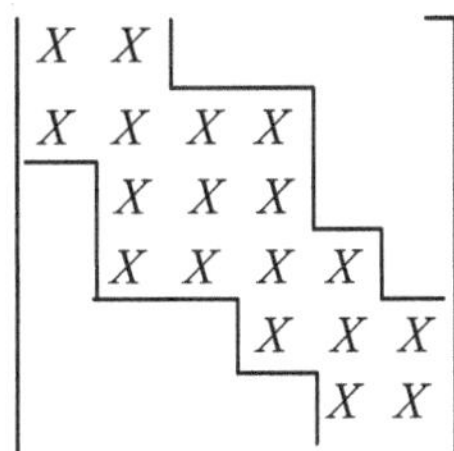

Figure 4. The graph underlying $\mathbf{A}$ is a tree.

$\mathbf{x} \in R^n$. The stiffness matrix $\mathbf{K}$ of an (undamped) anchored system is PD; $\mathbf{x}^T\mathbf{K}\mathbf{x}$ is related to the potential, or strain, energy of the system. It is well-known that $\mathbf{x}^T\mathbf{A}\mathbf{x} > 0$ if and only if all the *principal minors* of $\mathbf{A}$ are positive. A principal minor of $\mathbf{A}$ is a minor taken from a set of columns and the *same* set of rows. $\mathbf{A} \in S_n$ is PD if and only if its eigenvalues $(\lambda_i)_1^n$ are all *positive*.

Now consider the concept of a *positive* matrix. If $\mathbf{A} \in M_n$, then

$\mathbf{A} \geq \mathbf{0}$	means that each entry in $\mathbf{A}$ is *non-negative*
$\mathbf{A} > \mathbf{0}$	means $\mathbf{A} \geq \mathbf{0}$ and at least one entry is *positive*
$\mathbf{A} >> \mathbf{0}$	means that *each* entry in $\mathbf{A}$ is *positive.*

Let

$$\mathcal{Z}_n = \{\mathbf{A} : \mathbf{A} \in M_n; \quad a_{ij} \leq 0, \quad i \neq j\}. \tag{20}$$

If $\mathbf{A} \in Z_n$, then it may be expressed in the form

$$\mathbf{A} = s\mathbf{I} - \mathbf{B}, \quad s > 0, \quad \mathbf{B} \geq \mathbf{0}. \tag{21}$$

The spectral radius of $\mathbf{B} \in M_n$ is

$$\rho(\mathbf{B}) = \max\{|\lambda|; \quad \lambda \text{ is an eigenvalue of } \mathbf{B}\}.$$

A matrix $\mathbf{A} \in M_n$ of the form (21) with $s \geq \rho(\mathbf{B})$ is called an *M-matrix*; if $s > \rho(\mathbf{B})$ it is a *non-singular* M-matrix.

Berman and Plemmons (1994) construct an inference tree of properties of non-singular M-matrices; one of the most important properties is that $\mathbf{A}^{-1} > \mathbf{0}$; each entry in $\mathbf{A}^{-1}$ is non-negative and at least one is positive.

A symmetric non-singular M-matrix is called a *Stieltjes* matrix, and importantly $\mathbf{A} \in S_n \cap Z_n$ is a Stieltjes matrix iff it is PD. Moreover, if $\mathbf{A}$

is an irreducible Stieltjes matrix then $\mathbf{A}^{-1} >> \mathbf{0}$: each entry in $\mathbf{A}^{-1}$ is strictly positive. This result has important consequences for some physical systems. For a physical system, like an on-line spring mass system, or a finite element model of a membrane in out-of-plane movement, the relation between static displacement and applied force has the form

$$\mathbf{Ku} = \mathbf{f}.$$

For an anchored spring-mass system, the stiffness matrix is an irreducible non-singular M-matrix, so that $\mathbf{K}^{-1} >> 0$. This agrees with our intuition that a positive force applied to any mass will displace each mass in the positive direction. Gladwell et al. (2009) showed that the analogue of this result holds for a FE model of a membrane.

For in-line spring-mass systems, and for the model of a beam in flexure, the stiffness matrix satisfies more stringent conditions which ensure that the eigenvalues are distinct, that the first eigenmode has no node, that the second mode has one node, etc. The matrix property is so-called *total positivity.* The definitions are as follows, and they refer to $\mathbf{A} \in M_n$, not just to $\mathbf{A} \in S_n$. A matrix $\mathbf{A} \in M_n$ is said to be

TN (totally non-negative) if all its minors are *non-negative*,
NTN if it is *non-singular* and TN,
TP (totally positive) if all its minors are *positive*,
O (oscillatory) if it is TN, and a power $\mathbf{A}^m$ is TP.

To these, we add one more: let $\mathbf{Z} = diag(+1, -1, +1, \ldots, (-)^{n-1})$, then $\tilde{\mathbf{A}} = \mathbf{ZAZ}$ is called the *sign-reverse* of $\mathbf{A}$.

SO (sign-oscillatory) if $\tilde{\mathbf{A}}$ is O.

Clearly, if $\mathbf{A} \in S_n$ is TP, then it is PD, but it can be shown that if $\mathbf{A} \in S_n$ is only NTN it is PD.

If $\mathbf{A} \in M_n$ then $\mathbf{A}$ is NTN, TP or O iff $(\tilde{\mathbf{A}})^{-1}$ is NTN, TP or O. Moreover, if $\mathbf{A}$ is SO, then $\mathbf{A}^{-1}$ is O, and vice versa.

Let us apply these results. The stiffness matrix $\mathbf{K}$ of an in-line spring mass system is a PD Jacobi matrix with *negative* off-diagonals; its *sign reverse* $\tilde{\mathbf{K}}$ is a PD Jacobi matrix with *positive* off-diagonal entries. The principal minors of $\tilde{\mathbf{K}}$ are positive, and it can be shown that its other minors are non-negative. If we form powers, $\tilde{\mathbf{K}}^2, \tilde{\mathbf{K}}^3, \ldots, \tilde{\mathbf{K}}^{n-1}$, then $\tilde{\mathbf{K}}^{n-1}$ is full. It can be shown that $\tilde{\mathbf{K}}^{n-1}$ is PD, so that $\tilde{\mathbf{K}}$ is O, and $\mathbf{K}$ is SO, and $\mathbf{K}^{-1}$ is O.

Total positivity is linked to staircase form through the result: If $\mathbf{A} \in S_n$ is NTN, it is a staircase matrix.

Gladwell (1998) proved that if $\mathbf{A} \in S_n$, if P denotes one of the properties NTN, O, TP, and if $\mathbf{B}$ is related to $\mathbf{A}$ by equations (14), (15) with μ not an eigenvalue of $\mathbf{A}$, then $\mathbf{B}$ has property P iff $\mathbf{A}$ has property P. This shows that these properties NTN, O, TP are *invariant* under $\mathbf{QR}$ factorization and reversal; in turn, this means that the minors of $\mathbf{B}$ will have the same sign properties as the minors of $\mathbf{A}$.

4 Lecture 4. Toda Flow

4.1 The Basic Equations

If $\mathbf{A}, \mathbf{B} \in M_n$, we denote the *Lie bracket* of $\mathbf{A}, \mathbf{B}$ by

$$[\mathbf{A}, \mathbf{B}] = \mathbf{AB} - \mathbf{BA}.$$

We note that if $\mathbf{A} \in S_n$, and $\mathbf{B} \in K_n$, i.e., $\mathbf{A}$ is symmetric and $\mathbf{B}$ is skew-symmetric, then $[\mathbf{A}, \mathbf{B}] \in S_n$, i.e., $[\mathbf{A}, \mathbf{B}]$ is symmetric.

We recall a basic result from Nanda (1985).

Let $\mathbf{S}(t) \in K_n$ be a skew symmetric matrix defined on $-\infty < t < \infty$, i.e., $\mathbf{S}^T = -\mathbf{S}$. Let $\mathbf{Q}(t)$ be the solution of

$$\frac{d\mathbf{Q}}{dt} = -\mathbf{SQ}, \quad \mathbf{Q}(0) = \mathbf{I}, \tag{22}$$

then (i) $\mathbf{Q}(t)$ is an orthogonal matrix for $-\infty < t < \infty$.

Let $\mathbf{A}(t)$ be the solution of

$$\frac{d\mathbf{A}}{dt} = [\mathbf{A}, \mathbf{S}], \quad \mathbf{A}(0) = \mathbf{A}_0 \in S, \tag{23}$$

then (ii)

$$\mathbf{A}(t) = \mathbf{Q}(t)\mathbf{A}_0\mathbf{Q}^T(t),$$

where $\mathbf{Q}$ is the solution of (22).

Proof (i)

$$\begin{aligned} \tfrac{d}{dt}(\mathbf{Q}^T\mathbf{Q}) &= \tfrac{d\mathbf{Q}^T}{dt}\mathbf{Q} + \mathbf{Q}^T\tfrac{d\mathbf{Q}}{dt}, \\ &= -\mathbf{Q}^T\mathbf{S}^T\mathbf{Q} - \mathbf{Q}^T\mathbf{SQ}, \\ &= -\mathbf{Q}^T(\mathbf{S}^T + \mathbf{S})\mathbf{Q} = \mathbf{0}. \end{aligned}$$

Therefore

$$\mathbf{Q}^T(t)\mathbf{Q}(t) = \text{constant} = \mathbf{Q}^T(0)\mathbf{Q}(0) = \mathbf{I} :$$

$\mathbf{Q}(t)$ is an orthogonal matrix.

(ii)

$$\frac{d}{dt}(\mathbf{Q}^T\mathbf{AQ}) = -(\mathbf{Q}^T\mathbf{S}^T)\mathbf{AQ} + \mathbf{Q}^T(\mathbf{AS} - \mathbf{SA})\mathbf{Q} + \mathbf{Q}^T\mathbf{A}(-\mathbf{SQ}) = \mathbf{0}.$$

Hence $\mathbf{Q}^T\mathbf{A}\mathbf{Q} = const = \mathbf{Q}^T(0)\mathbf{A}_0\mathbf{Q}(0) = \mathbf{A}_0$, and

$$\mathbf{A}(t) = \mathbf{Q}\mathbf{A}_0\mathbf{Q}^T.$$

The flow of $\mathbf{A}(t)$ according to (23), i.e.,

$$\dot{\mathbf{A}} = \mathbf{A}\mathbf{S} - \mathbf{S}\mathbf{A}, \tag{24}$$

where $\mathbf{S}^T = -\mathbf{S}$ is an orthogonal matrix, is an *isospectral* flow, called a *Toda* Flow. The analysis had its beginning in the investigation of the so-called Toda lattice, Toda (1970). See Chu (1984) or Gladwell (2004) for a review of the literature.

Historically, Toda flow was applied first to a Jacobi matrix. Suppose

$$\mathbf{A} = \begin{bmatrix} a_1 & b_1 & & & \\ b_1 & a_2 & b_2 & & \\ & \ddots & \ddots & \ddots & \\ & & \ddots & \ddots & b_{n-1} \\ & & & b_{n-1} & a_n \end{bmatrix},$$

$$\mathbf{S} = \begin{bmatrix} 0 & b_1 & & & \\ -b_1 & 0 & b_2 & & \\ & -b_2 & \ddots & \ddots & \\ & & \ddots & \ddots & b_{n-1} \\ & & & -b_n & 0 \end{bmatrix}.$$

Then it can easily be verified that, as it flows, $\mathbf{A}$ remains a Jacobi matrix, i.e., if $|i-j| > 1$ then $\dot{a}_{ij} = 0$. Since $a_{ij}(0) = 0$ if $|i-j| > 1$ then $a_{ij}(t) = 0$ for all t. The off-diagonal entries b_i satisfy

$$\dot{b}_i = (a_i - a_{i+1})b_i \qquad i = 1, 2, \ldots, n-1,$$

which implies that b_i retains its sign: if $b_i(0) > 0$ then $b_i(t) > 0$.

We conclude that in this simple example, the matrix $\mathbf{A}$ retains both its zero/non-zero structure and its sign properties. We can therefore use Toda flow to construct spring-mass systems which vary over 'time' so that their eigenvalues remain the same. We use the Toda flow to find $\mathbf{A}(t)$, and then decompose $\mathbf{A}$ into $\mathbf{M}$ and $\mathbf{K}$ as described in Section 1.3.

There is a little-known Toda flow that keeps *periodic* Jacobi form invariant:

$$\mathbf{A} = \begin{bmatrix} a_1 & b_1 & & & b_n \\ b_1 & a_2 & b_2 & & \\ & \ddots & \ddots & \ddots & \\ & & \ddots & \ddots & b_{n-1} \\ b_n & & & b_{n-1} & a_n \end{bmatrix},$$

$$\mathbf{S} = \begin{bmatrix} 0 & b_1 & & & -b_n \\ -b_1 & 0 & b_2 & & \\ & -b_2 & \ddots & \ddots & \\ & & \ddots & \ddots & b_{n-1} \\ b_n & & & -b_{n-1} & 0 \end{bmatrix}.$$

We noted earlier that a periodic Jacobi matrix is somehow like a Jacobi matrix inscribed on a cylinder: b_n appears to the left of a_1 and to the right of a_n. In $\mathbf{S}$, the entry b_n on the first row is treated as if it were in the lower triangle, and assigned negative sign; similarly the entry b_n on the last row is treated as if it were in the upper triangle, and given a positive sign.

We conclude that Toda flow of a periodic Jacobi matrix is no more complicated than flow of a Jacobi matrix; this contrasts with the difficulty of the algebraic solution to the inverse eigenvalue problem for the periodic Jacobi matrix, as described in Section 2.4.

4.2 Application of Toda Flow

If $\mathbf{A} \in S_n$ then we define $\mathbf{A}^+$ as the upper triangle of $\mathbf{A}$. If $\mathbf{S} = \mathbf{A}^+ - \mathbf{A}^{+T}$, then $\mathbf{S}$ is skew symmetric. Now we can show that if $\mathbf{A}(0)$ is a staircase matrix, then so is $\mathbf{A}(t)$, the solution of (23), with the same staircase form as $\mathbf{A}(0)$. Let P denote one of the properties TN, NTN, TP, O, SO. If $\mathbf{A}(0) \in S_n$ has property P then $\mathbf{A}(t)$, the solution of (23) with $\mathbf{S} = \mathbf{A}^+ - \mathbf{A}^{+T}$, has the same property P: all these properties are *invariant* under Toda flow.

While these are interesting and important properties of Toda flow, they apply only to staircase matrices, and such matrices appear only when the physical system is in a certain sense an in-line system, such as a rod or a beam: its underlying graph is a generalized path. To find a skew symmetric $\mathbf{S}$ that ensures that, as it flows, $\mathbf{A}(t)$ lies on a certain graph, we must modify the concept of Toda flow.

In Section 2.5, we defined the associated graph, $\mathcal{G}(\mathbf{A})$, of $\mathbf{A} \in S_n$: $(P_i, P_j) \in \mathcal{E}$ iff $a_{ij} \neq 0$. We need a somewhat looser term: the matrix

$\mathbf{A} \in \mathcal{S}_n$ is said to *lie* on $\mathcal{G}$ if $a_{ij} = 0$ whenever $(P_i, P_j) \notin \mathcal{E}$; a_{ij} may be zero even when $(P_i, P_j) \in \mathcal{E}$.

Let $\mathbf{T} \in \mathcal{K}_n$ be defined by $\mathbf{T} = (t_{ij})$ where

$$t_{ij} = \begin{cases} +1 & \text{if} \quad i < j, \\ 0 & \text{if} \quad i = j, \\ -1 & \text{if} \quad i > j. \end{cases}$$

If $\mathbf{S} \in \mathcal{K}_n$, then $\mathbf{T} \circ \mathbf{S} \in \mathcal{S}_n$, where $\circ$ denotes the Hadamard (entry wise) product. If $\mathbf{S} \in \mathcal{K}_n$ and $\mathbf{T} \circ \mathbf{S}$ lies on $\mathcal{G}$, we will say that $\mathbf{S}$ *lies* on $\mathcal{G}$.

Given a graph $\mathcal{G}$, we may form its *complement* $\bar{\mathcal{G}}$; this is the graph on the same vertex set $\mathcal{V}(\mathcal{G})$, but is such that two vertices of $\bar{\mathcal{G}}$ are adjacent iff they are not adjacent in $\mathcal{G}$. If $\mathbf{A} \in \mathcal{S}_n$ and $\mathcal{G}$ is a graph with $\mathcal{V}(\mathcal{G}) = \{P_1, P_2, \ldots, P_n\}$ then we may partition $\mathbf{A}$ in the form $\mathbf{A} = \mathbf{A}_1 + \mathbf{A}_2$, where $\mathbf{A}_1$ lies on $\mathcal{G}$ and $\mathbf{A}_2$ lies on $\bar{\mathcal{G}}$. Thus, the subscript 1 denotes the restriction of $\mathbf{A}$ to $\mathcal{G}$, and the subscript 2 denotes the restriction to $\bar{\mathcal{G}}$, the complement of $\mathcal{G}$. For definiteness, we place the diagonal entries of $\mathbf{A}$ in $\mathbf{A}_1$. Similarly, if $\mathbf{S} \in \mathcal{K}_n$ we may partition $\mathbf{S}$ in the form $\mathbf{S} = \mathbf{S}_1 + \mathbf{S}_2$, where $\mathbf{S}_1$ lies on $\mathcal{G}$ and $\mathbf{S}_2$ lies on $\bar{\mathcal{G}}$.

Now return to equation (24). If $\mathbf{A}(0)$ lies on a graph $\mathcal{G}$, then $\mathbf{A}(t)$, governed by (24), will not in general remain on $\mathcal{G}$; we must choose $\mathbf{S} \in \mathcal{K}_n$ to constrain $\mathbf{A}(t)$ to remain on $\mathcal{G}$. We may write equation (24) as

$$\dot{\mathbf{A}}_1 + \dot{\mathbf{A}}_2 = [\mathbf{A}_1 + \mathbf{A}_2, \mathbf{S}] = [\mathbf{A}_1, \mathbf{S}] + [\mathbf{A}_2, \mathbf{S}]. \tag{25}$$

Now choose $\mathbf{S}$ so that $[\mathbf{A}_1, \mathbf{S}]$ lies on $\mathcal{G}$, i.e.,

$$[\mathbf{A}_1, \mathbf{S}]_2 = \mathbf{0}. \tag{26}$$

Since $\mathbf{A}(0)$ is on $\mathcal{G}$, so that $\mathbf{A}_2(0) = \mathbf{0}$, equations (25), (26) show that

$$\dot{\mathbf{A}}_2 = [\mathbf{A}_2, \mathbf{S}]_2, \quad \mathbf{A}_2(0) = \mathbf{0}$$

which has the unique solution $\mathbf{A}_2(t) = 0 : \mathbf{A}(t)$ is on $\mathcal{G}$, i.e., $\mathbf{A}(t) = \mathbf{A}_1(t)$, and

$$\dot{\mathbf{A}}_1 = [\mathbf{A}_1, \mathbf{S}]_1. \tag{27}$$

We note that the equations governing the isospectral flow of $\mathbf{A}_1(t)$ have been reduced to the differential equation (27) on $\mathcal{G}$, and the simultaneous linear algebraic equations, (26), i.e.,

$$[\mathbf{A}_1, \mathbf{S}_1]_2 + [\mathbf{A}_1, \mathbf{S}_2]_2 = \mathbf{0},$$

for $\mathbf{S}$ on $\bar{\mathcal{G}}$. In general, $\mathbf{S}$ will not lie on $\mathcal{G}$, i.e., $\mathbf{S}_2 \neq \mathbf{0}$. What we found earlier, for the Jacobi, and staircase matrices was that there was a solution, namely

$$\mathbf{S} = \mathbf{T0A}$$

which lay on $\mathcal{G}$. We found also that there was a solution $\mathbf{S}$ for a periodic Jacobi matrix which lay on $\mathcal{G}$. Gladwell and Rojo (2007) showed that the $\mathbf{S}$ for a generalized Jacobi matrix could be generalized to give $\mathbf{S}$ on $\mathcal{G}$ for a so-called *periodic staircase* matrix.

In general, $\mathbf{S}$ will not lie on $\mathcal{G}$. In that case we take $\mathbf{S}_1 = \mathbf{T0A}_1$, find $\mathbf{S}_2$ so that (26) holds, and then find $\mathbf{A}_1$ from equation (27). Gladwell et al. (2009) implement this procedure to find an isospectral flow of a finite element model of a membrane.

Bibliography

S.A. Andrea and T.G. Berry. Continued fractions and periodic Jacobi matrices. *Linear Algebra and Its Applications*, 161:117–134, 1992.

P. Arbenz and G.H. Golub. Matrix shapes invariant under the symmetric QR algorithm. *Numerical Linear Algebra with Applications*, 2:87–93, 1995.

A. Berman and R.J. Plemmons. *Nonnegative matrices in the mathematical sciences.* Society for Industrial and Applied Mathematics, 1994.

D. Boley and G.H. Golub. A modified method for reconstructing periodic Jacobi matrices. *Mathematics of Computation*, 42:143–150, 1984.

D. Boley and G.H. Golub. A survey of matrix inverse eigenvalue problems. *Inverse Problems*, 3:595–622, 1987.

M.T. Chu. The generalized Toda flow, the QR algorithm and the center manifold theory. *SIAM Journal on Algebraic and Discrete Methods*, 5: 187–201, 1984.

M.T. Chu and G.H. Golub. *Inverse Eigenvalue Problems: Theory, Algorithms and Applications.* Oxford University Press, 2005.

A.L. Duarte. Construction of acyclic matrices from spectral data. *Linear Algebra and Its Applications*, 113:173–182, 1989.

S. Friedland. Inverse eigenvalue problems. *Linear Algebra and Its Applications*, 17:15–51, 1977.

G.M.L. Gladwell. *Inverse Problems in Vibration.* Kluwer Academic Publishers, 2004.

G.M.L. Gladwell. Minimal mass solutions to inverse eigenvalue problems. *Inverse Problems*, 22:539–551, 2006.

G.M.L. Gladwell. The inverse problem for the vibrating beam. *Proceedings of the Royal Society of London - Series A*, 393:277–295, 1984.

G.M.L. Gladwell. Total positivity and the QR algorithm. *Linear Algebra and Its Applications*, 271:257–272, 1998.

G.M.L. Gladwell and O. Rojo. Isospectral flows that preserve matrix structure. *Linear Algebra and Its Applications*, 421:85–96, 2007.

G.M.L. Gladwell, K.R. Lawrence, and D. Siegel. Isospectral finite element membranes. *Mechanical Systems and Signal Processing*, 23:1986–1999, 2009.

G.H. Golub. Some uses of the Lanczos algorithm in numerical linear algebra. In J.H.H. Miller, editor, *Topics in Numerical Analysis*. Academic Press, 1973.

T. Nanda. Differential equations and the QR algorithm. *SIAM Journal of Numerical Analysis*, 22:310–321, 1985.

P. Nylen and F. Uhlig. Realizations of interlacing by tree-patterned matrices. *Linear Multilinear Algebra*, 38:13–37, 1994.

Lord Rayleigh. *The Theory of Sound*. Macmillan, 1894.

M. Toda. Waves in nonlinear lattices. *Progress of Theoretical Physicis Supplement*, 45:174–200, 1970.

D.K. Vijay. *Some Inverse Problems in Mechanics*. University of Waterloo, M.A.Sc. Thesis, 1972.

S.F. Xu. *An Introduction to Inverse Algebraic Eigenvalue Problems*. Vieweg, 1998.

An Introduction to Classical Inverse Eigenvalue Problems

Antonino Morassi

Via Bortulin 9, Liariis di Ovaro (Ud), Italy

Abstract Aim of these notes is to present an elementary introduction to classical inverse eigenvalue problems in one dimension. Attention is mainly focused on Sturm-Liouville differential operators given in canonical form on a finite interval. A uniqueness result for a fourth order Euler-Bernoulli operator is also discussed.

1 Introduction

Eigenvalue inverse problems are primarily model problems simulating situations of practical interest. Typically, in an inverse eigenvalue problem, one measures a certain number of natural frequencies of a vibrating system and tries to infer some physical properties of the system under observation.

To fix ideas, consider a thin straight rod of length L, made by homogeneous, isotropic linear elastic material with constant Young's modulus E, $E > 0$, and constant volume mass density γ, $\gamma > 0$. The free vibrations of the rod are governed by the partial differential equation

$$\frac{\partial}{\partial x}\left(EA(x)\frac{\partial w(x,t)}{\partial x}\right) - \gamma A(x)\frac{\partial^2 w(x,t)}{\partial^2 t} = 0, \quad (x,t) \in (0,L) \times (0,\infty), \quad (1)$$

where $A = A(x)$ is the area of the cross section and $w = w(x,t)$ is the longitudinal displacement of the cross section of abscissa x evaluated at the time t. The function $A = A(x)$ is assumed to be regular (i.e., $C^2([0,L])$) and strictly positive in $[0,L]$. The end cross-sections of the rod are assumed to be fixed, namely

$$w(0,t) = 0 = w(L,t), \quad t \geq 0. \quad (2)$$

If the rod vibrates with frequency ω and spatial shape $X = X(x)$, i.e. $w(x,t) = X(x)\cos(\omega t)$, then the free vibrations are governed by the boundary value problem

$$\begin{cases} (A(x)X'(x))' + \lambda A(x)X(x) = 0, & \text{in } (0,L), \quad (3)\\ X(0) = 0 = X(L), & \quad (4) \end{cases}$$

where $\lambda = \frac{E}{\gamma}\omega^2$ and $X \in C^2([0,L]) \setminus \{0\}$. The (real) number λ is called eigenvalue of the rod and the corresponding function $X = X(x)$ is the eigenfunction associated to λ. The pair $\{\lambda, X(x)\}$ is a Dirichlet eigenpair. For example, if $A(x) \equiv const.$ in $[0,L]$, then $\lambda_n = \left(\frac{n\pi}{L}\right)^2$ and $X_n(x) = \sqrt{\frac{2}{L}} \sin \frac{n\pi x}{L}$, $n \geq 1$.

We find convenient to rewrite the problem (3)-(4) in the Sturm-Liouville canonical form by putting

$$y(x) = \sqrt{A(x)}X(x). \tag{5}$$

Then, the new eigenpair $\{\lambda, y(x)\}$ solves

$$\begin{cases} y''(x) + \lambda y(x) = q(x)y(x), & \text{in } (0,1), \quad (6)\\ y(0) = 0 = y(1), & \quad (7) \end{cases}$$

where the potential $q = q(x)$ is defined as

$$q(x) = \frac{(\sqrt{A(x)})''}{\sqrt{A(x)}}, \quad \text{in } (0,1), \tag{8}$$

and where, to simplify the notation, we have assumed $L = 1$.

The *direct* eigenvalue problem consists in finding the eigenvalues $\{\lambda_n\}_{n=1}^{\infty}$ and eigenfunctions $\{y_n(x)\}_{n=1}^{\infty}$ of (6)-(7) for a potential $q = q(x)$ given in $[0,1]$ or, equivalently, for a given cross section $A = A(x)$ of the rod.

Conversely, the *inverse* eigenvalue problem consists, for example, in finding information on the potential $q = q(x)$ in $[0,1]$ given the Dirichlet spectrum $\{\lambda_n\}_{n=1}^{\infty}$ of the rod.

Inverse eigenvalue problems for Sturm-Liouville operators $Ly = -y'' + qy$ on a finite interval can be considered as a rather consolidated topic in the literature of dynamical inverse problems. Fundamental contributions were given in this field by Borg, Levinson, Krein, Gel'fand and Levitan, Marchenko, Hochstadt, Hald, McLaughlin, Trubowitz and others, see the books by Levitan (1987), Pöschel and Trubowitz (1987) and Gladwell (2004) for a comprehensive presentation.

Aim of these lectures is to present an elementary introduction to these inverse problems. In the first lecture, some general properties of the direct eigenvalue problem and asymptotic estimates of eigenvalues and eigenfunctions are recalled. The second lecture is devoted to the presentation of some classical uniqueness results obtained by Borg (1946). The third lecture illustrates the approach by Hochstadt (1973) in the absence of data sufficient to ensure the uniqueness of the inverse problem, and a first stability result. A local existence result and a reconstruction procedure, still due to Borg (1946), are discussed in the fourth lecture. Finally, the fifth lecture is devoted to an inverse problem for the Euler-Bernoulli operator of the bending vibration of a beam given by Elcrat and Papanicolaou (1997).

In all the lectures I plan to give the general ideas of the methods instead of the exact proofs. These were included in the notes delivered during the course and, in any case, can be tracked down from the original papers cited in the references. The presentation of the arguments is almost self-contained and prerequisites are basic knowledge of functional analysis (Brezis (1986), Friedman (1982)) and complex analysis (Ahlfors (1984), Titchmarsh (1939)).

In selecting the topics for presentation I had to take into account the limited time available for a rather extensive subject. I was forced to sacrifice some arguments (for example, the theory by Gel'fand and Levitan (1955), see also Levitan (1968)) that would have merited more attention, and to omit some other topics of current interest, such as, for example, isospectral systems (Pöschel and Trubowitz (1987), Coleman and McLaughlin (1993) and Gladwell and Morassi (1995)) and inverse nodal problems (Hald and McLaughlin (1989)).

2 Properties of the Dirichlet Eigenvalue Problem

We begin by recalling some basic properties of the Dirichlet eigenvalue problem (6)-(7) for real-valued square summable potential q on $(0,1)$, i.e. $q \in L^2(0,1)$.

i) There exists a sequence of Dirichlet eigenvalues $\{\lambda_n\}_{n=1}^{\infty}$. The eigenvalues are real numbers and $\lim_{n\to\infty} \lambda_n = +\infty$.

ii) The eigenvalues are simple, that is

$$\lambda_1 < \lambda_2 < ... < \lambda_n < ... \tag{9}$$

and the eigenspace $\mathcal{U}_n$ associated to the nth eigenvalue λ_n, $n \geq 1$, is given by

$$\mathcal{U}_n = \text{span}\{g_n\}, \tag{10}$$

where $g_n = g_n(x)$ is the eigenfunction associated to λ_n satisfying the normalization condition $\int_0^1 g_n^2(x)dx = 1$.

iii) The family $\{g_n\}_{n=1}^{\infty}$ is an orthonormal basis of the space $\mathcal{D}$ of continuous functions which vanish at $x = 0$ and $x = 1$, that is:

$$\int_0^1 g_n(x)g_m(x)dx = \delta_{nm} = \begin{cases} 1 & \text{if } n = m, \\ 0 & \text{if } n \neq m, \quad n, m \geq 1, \end{cases} \tag{11}$$

and for every $f \in \mathcal{D}$ the series

$$\sum_{n=1}^{\infty} c_n g_n(x), \quad \text{with } c_n = \int_0^1 f(x)g_n(x)dx, \tag{12}$$

converges uniformly to f in $[0, 1]$.

The above properties can be deduced from abstract spectral theory for self-adjoint compact operators defined on Hilbert spaces, see Brezis (1986) and Friedman (1982). An alternative approach is based on functional theoretical methods and specific properties of the Sturm-Liouville operators, see Titchmarsh (1962).

As we will see in the next sections, the asymptotic behavior of eigenpairs plays an important role in inverse spectral theory. With reference to the Dirichlet problem (6)-(7), let us consider the solution $y = y(x, \lambda)$ to the initial value problem

$$\begin{cases} y''(x) + \lambda y(x) = q(x)y(x), \quad \text{in } (0, 1), & (13) \\ y(0) = 0, & (14) \\ y'(0) = 1, & (15) \end{cases}$$

for some (possibly complex) number λ and for a (possibly complex-valued) L^2 potential q. By considering the right hand side as a forcing term, y can be interpreted as the displacement of an harmonic oscillator with frequency $\sqrt{\lambda}$. Then, by the Duhamel's representation, the function y is the solution of the Volterra linear integral equation

$$y(x, \lambda) = \frac{\sin\sqrt{\lambda}x}{\sqrt{\lambda}} + \frac{1}{\sqrt{\lambda}} \int_0^x \sin\sqrt{\lambda}(x - t)q(t)y(t)dt, \quad x \in [0, 1]. \tag{16}$$

It can be shown that there exists a unique solution belonging to $C^1([0, 1])$ of (16) ($y \in C^2([0, 1])$ if q is continuous). Moreover, y is an entire function of order $\frac{1}{2}$ in λ. We recall that a function $f = f(\lambda)$ of the complex variable λ is an analytic function if has the derivative whenever f is defined. If f is analytic in the whole plane, then f is said to be an entire function. Let $M(r) = \max_{|\lambda|=r} |f(\lambda)|$. An entire function $f = f(\lambda)$ has order s if s is the smallest number such that $M(r) \leq \exp(r^{s+\epsilon})$ for any given $\epsilon > 0$, as $r \to \infty$.

By (16) one can also deduce that $y(x,\lambda) \approx \frac{\sin\sqrt{\lambda}x}{\sqrt{\lambda}}$ for large $|\lambda|$, precisely

$$\left| y(x,\lambda) - \frac{\sin\sqrt{\lambda}x}{\sqrt{\lambda}} \right| \geq \frac{\|q\|_{L^2}}{|\lambda|} \exp\left(x(\|q\|_{L^2} + |Im\sqrt{\lambda}|) \right), \tag{17}$$

uniformly in $[0,1]$, where $\|q\|_{L^2} = \left(\int_0^1 |q(x)|^2 dx \right)^{1/2}$, $\sqrt{\lambda} = Re\sqrt{\lambda} + iIm\sqrt{\lambda}$, $i = \sqrt{-1}$. The zeros of $y(1,\lambda) = 0$ are the eigenvalues $\{\lambda_n\}$ of the Dirichlet problem (6)-(7) and, in this case, $y = y(x,\lambda_n)$ is the associated eigenfunction. By estimate (17), we expect that higher order zeros of $y(1,\lambda) = 0$ are close to the (square of) zeros of $\sin\sqrt{\lambda} = 0$, that is $\lambda_n \approx (n\pi)^2$ as $n \to \infty$. More precisely, consider the circle $C_n = \{z \in \mathbb{C} |\ |z - n\pi| = \frac{\pi}{4}\}$ for n large enough. One can prove that there exists $N \in \mathbb{N}$ such that for every $n \geq N$ there exists exactly one zero of $y(1,\lambda) = 0$ inside the circle C_n, namely the following eigenvalue asymptotic estimate holds

$$|\sqrt{\lambda} - n\pi| < \frac{\pi}{4}. \tag{18}$$

The proof of (18) is based on a well-known result of complex analysis: the Rouché's Theorem, see Ahlfors (1984). Let $f = f(z)$, $g = g(z)$ be two analytic functions inside C_n and assume that $|f(z)| < |g(z)|$ on C_n. Then, Rouché's Theorem states that $g(z)$ and $g(z) + f(z)$ have exactly the same number of zeros inside C_n. In order to apply this theorem, let us formally rewrite the function $y(1,\lambda)$ as

$$y(1,\lambda) = \underbrace{\left(y(\lambda) - \frac{\sin\sqrt{\lambda}x}{\sqrt{\lambda}} \right)}_{\equiv f(\lambda)} + \underbrace{\frac{\sin\sqrt{\lambda}x}{\sqrt{\lambda}}}_{\equiv g(\lambda)} \tag{19}$$

The delicate point consists in showing that $|f(\lambda)| < |g(\lambda)|$ on C_n. Then, recalling that $g(\lambda)$ has exactly one zero $(\sqrt{\tilde{\lambda}_n} = n\pi)$ inside C_n, one obtains (18).

Now, inserting the eigenvalue asymptotic estimate (18) in estimate (17) we get

$$y(x,\lambda_n) = \frac{\sin\sqrt{\lambda_n}x}{\sqrt{\lambda_n}} + O\left(\frac{1}{n^2}\right), \quad \text{as } n \to \infty. \tag{20}$$

Recalling that $g_n(x) = \frac{y(x,\lambda_n}{\|y(x,\lambda_n)\|_{L^2}}$ we obtain the asymptotic eigenfunction estimate

$$g_n(x) = \sqrt{2}\sin(n\pi x) + O\left(\frac{1}{n}\right), \tag{21}$$

which holds uniformly on bounded subsets of $[0,1] \times L^2(0,1)$ as $n \to \infty$. Finally, by iterating the above procedure, the eigenvalue estimate (18) can be improved to obtain

$$\lambda_n = (n\pi)^2 + \int_0^1 q(x)dx - \int_0^1 \cos(2n\pi x)q(x)dx + O\left(\frac{1}{n}\right), \quad \text{as } n \to \infty. \tag{22}$$

3 Uniqueness Results: the Borg's Approach

3.1 Symmetric Potential and Dirichlet Boundary Conditions

Let us consider the Dirichlet eigenvalue problem

$$\begin{cases} y''(x) + \lambda y(x) = q(x)y(x), & \text{in } (0,1), & (23)\\ y(0) = 0 = y(1), & & (24) \end{cases}$$

where $q \in L^2(0,1)$ is a real-valued potential. Denote by $\{g_n(x,q), \lambda_n\}_{n=1}^{\infty}$, $\int_0^1 g_n^2 dx = 1$, the eigenpairs of (23)-(24). Let us compare the high order eigenvalues of the above problem with those of the *reference* problem with $q \equiv 0$:

$$\begin{cases} z''(x) + \chi z(x) = 0, & \text{in } (0,1), & (25)\\ z(0) = 0 = z(1), & & (26) \end{cases}$$

An easy calculation shows that $\chi_n = (n\pi)^2$, $g_n(x,0) = \sqrt{2}\sin(n\pi x)$, $n \geq 1$. Then, by the asymptotic eigenvalue estimate (22) we obtain

$$\lambda_n = \chi_n + \int_0^1 q(x)dx - \int_0^1 \cos(2n\pi x)q(x)dx + O\left(\frac{1}{n}\right), \quad \text{as } n \to \infty. \tag{27}$$

Generally speaking, it turns out that knowledge of the high order eigenvalues can give information *only* on the average value of q and on the higher order Fourier coefficients of q evaluated on the family $\{\cos(2n\pi x)\}_{n=1}^{\infty}$. Since the family $\{\sqrt{2}\cos(2n\pi x)\}_{n=1}^{\infty} \cup \{1\}$ is an orthonormal basis of the space of the even functions with respect to $x = \frac{1}{2}$

$$L^2_{even}(0,1) = \{f \in L^2(0,1) |\ f(x) = f(1-x) \text{ a.e. in } (0,1)\}, \tag{28}$$

we expect to be able only to extract information from $\{\lambda_n\}_{n=1}^{\infty}$ for the even part of the potential q.

These heuristic considerations were made rigorous in the following celebrated theorem by Borg (1946).

Theorem 3.1 (Borg (1946)). *Let $q \in L^2_{even}(0,1)$. The potential q is uniquely determined by the full Dirichlet spectrum $\{\lambda_n\}_{n=1}^{\infty}$.*

The proof is by contradiction. Suppose that there exist another potential $p \in L^2_{even}(0,1)$, $p \neq q$, such that the eigenvalue problem

$$\begin{cases} z''(x) + \lambda z(x) = p(x)z(x), \quad \text{in } (0,1), & (29) \\ z(0) = 0 = z(1), & (30) \end{cases}$$

has exactly the same eigenvalues of (23)-(24), i.e. $\lambda_n(p) = \lambda_n(q)$ for every $n \geq 1$. In order to simplify the notation, let us denote by $g_n(q) = g_n(x,q)$, $g_n(p) = g_n(x,p)$ the normalized Dirichlet eigenfunctions associated to λ_n for potential q and p, respectively. Note that the nth Dirichlet eigenfunction is even when n is odd, and is odd when n is even. By multiplying the differential equation satisfied by $g_n(q)$, $g_n(p)$ by $g_n(p)$, $g_n(q)$, respectively, integrating by parts in $(0,1)$ and subtracting, we obtain

$$\int_0^1 (q-p)g_n(p)g_n(q)dx = 0, \quad \text{for every } n \geq 1. \tag{31}$$

By the asymptotic eigenvalue estimate (22) we have

$$\int_0^1 q dx = \int_0^1 p dx \tag{32}$$

and then condition (31) can be written as

$$\int_0^1 (q-p)(1 - g_n(p)g_n(q))dx = 0, \quad \text{for every } n \geq 1. \tag{33}$$

Therefore, to find the contradiction it is enough to show that the family $\{1\} \cup \{1 - g_n(q)g_n(p)\}_{n=1}^{\infty}$ is a *complete* system of functions in $L^2_{even}(0,1)$. Actually, we shall prove that this family is a basis of $L^2_{even}(0,1)$. We recall that a sequence of vectors $\{v_n\}_{n=1}^{\infty}$ in a separable Hilbert space H is a basis for H if there exists a Hilbert space h of sequences $\alpha = (\alpha_1, \alpha_2, ...)$ such that the correspondence $\alpha \to \sum_{n=1}^{\infty} \alpha_n v_n$ is a linear isomorphism between h and H (that is, an isomorphism which is continuous and has continuous inverse).

Let us introduce the set of functions $\{U_n\}_{n=0}^{\infty}$ defined as

$$\begin{cases} U_0(x) = 1, & (34) \\ U_n(x) = \sqrt{2}\left(\int_0^1 g_n(q)g_n(p)dx - g_n(q)g_n(p)\right), \quad n \geq 1. & (35) \end{cases}$$

Clearly, $\{U_n\}_{n=0}^{\infty}$ is a bounded subset of $L^2_{even}(0,1)$. We will prove that $\{U_n\}_{n=0}^{\infty}$ is a basis of $L^2_{even}(0,1)$. This immediately implies that also $\{1\} \cup \{g_n(q)g_n(p) - 1\}_{n=1}^{\infty}$ is a basis of $L^2_{even}(0,1)$. At this point we make use of the following useful result.

Theorem 3.2 (Pöschel and Trubowitz (1987), Theorem 3 of Appendix D). *Let $\{e_n\}_{n\geq 0}$ be an orthonormal basis of an Hilbert space H. Let $\{d_n\}_{n\geq 0}$ be a sequence of elements of H. If*

i) $\{d_n\}_{n=0}^{\infty}$ is such that $\sum_{n=0}^{\infty} \|e_n - d_n\|_H^2 < +\infty$,
and
ii) $\{d_n\}_{n=0}^{\infty}$ are linear independent in H,
then $\{d_n\}_{n=0}^{\infty}$ is a basis of H.

We recall that a sequence $\{v_n\}_{n=1}^{\infty}$ in a separable Hilbert space H is linearly independent if $\sum_{n=1}^{\infty} c_n v_n = 0$ for some sequence $\{c_n\}_{n=1}^{\infty}$ with $\sum_{n=1}^{\infty} c_n^2 < \infty$, then $c_n = 0$ for all n. We apply Theorem 3.2 with $H = L^2_{even}(0,1)$, $\{e_n\}_{n=0}^{\infty} = \{\sqrt{2}\cos(2n\pi x)\}_{n=1}^{\infty} \cup \{e_0 = 1\}$, $d_n = U_n$ for every $n \geq 0$.

Condition i) is easily checked. By the asymptotic eigenfunction estimate (21) we have

$$U_n(x) = \sqrt{2}\cos(2n\pi x) + O\left(\frac{1}{n}\right), \quad \text{as } n \to \infty \tag{36}$$

and therefore

$$\sum_{n\geq 0} \|e_n - U_n\|_{L^2}^2 = \sum_{n=1}^{\infty} O\left(\frac{1}{n^2}\right) < \infty. \tag{37}$$

The proof of the linear independence stated in condition ii) is more difficult. The original idea of Borg was to find a sequence of bounded functions $\{V_m\}_{m=0}^{\infty} \subset L^2_{even}(0,1)$ such that $\{U_n, V_m\}_{m,n=0}^{\infty}$ is a bi-orthonormal system of functions in $L^2_{even}(0,1)$, that is

$$\begin{cases} (U_n, V_n) = 1, & \text{for every } n \geq 0, \quad (38)\\ (U_n, V_m) = 0, & \text{for every } m, n \geq 0,\ n \neq m, \quad (39) \end{cases}$$

where $(U_n, V_m) = \int_0^1 U_n(x)V_m(x)dx$. Choose

$$\begin{cases} V_0(x) = 1, & (40)\\ V_m(x) = a_m'(x), \quad m \geq 1, & (41) \end{cases}$$

with

$$a_m(x) = g_m(x,q)\zeta_m(x,p), \tag{42}$$

where $\zeta_m = \zeta_m(x,p)$ is a *suitable* solution of the differential equation

$$\zeta_m'' + \lambda_m \zeta_m = p\zeta_m, \quad \text{in } (0,1). \tag{43}$$

Note that ζ_m is not an eigenfunction of p since ζ_m does not necessarily satisfy the Dirichlet boundary conditions at $x = 0$ and $x = 1$.

By definition, we have

$$(U_0, V_0) = 1, \quad (U_0, V_n) = 0, \quad (U_n, V_0) = 0, \quad n \geq 1. \tag{44}$$

Assume $m, n \geq 1$. Since

$$(U_n, V_m) = -\sqrt{2}(g_n(q)g_n(p), a_m') \tag{45}$$

and

$$(g_n(q)g_n(p), a_m') = -((g_n(q)g_n(p))', a_m), \tag{46}$$

a direct computation shows that

$$((g_n(q)g_n(p))', a_m) = \frac{1}{2}\int_0^1 \left[\zeta_m(p)g_n(p) \det \begin{pmatrix} g_m(q) & g_n(q) \\ g_m'(q) & g_n'(q) \end{pmatrix} + \right.$$
$$\left. + g_m(q)g_n(q) \det \begin{pmatrix} \zeta_m(p) & g_n(p) \\ \zeta_m'(p) & g_n'(p) \end{pmatrix}\right] dx. \tag{47}$$

If $m \neq n$, $m, n \geq 1$, then by using the differential equation satisfied by $g_m(q)$ and $\zeta_m(p)$ we have

$$(\lambda_m - \lambda_n)((g_n(q)g_n(p))', a_m) =$$
$$= \frac{1}{2} \det \begin{pmatrix} g_m(q) & g_n(q) \\ g_m'(q) & g_n'(q) \end{pmatrix} \det \begin{pmatrix} \zeta_m(p) & g_n(p) \\ \zeta_m'(p) & g_n'(p) \end{pmatrix} \Big|_{x=0}^{x=1} = 0, \tag{48}$$

since $g_m(x,q) = g_n(x,q) = 0$ at $x = 0$ and $x = 1$. This means that $(U_n, V_m) = 0$ for $m, n \geq 1$ and $m \neq n$.

Let $m = n$, $m, n \geq 1$. Recalling that

$$\det \begin{pmatrix} \zeta_n(p) & g_n(p) \\ \zeta_n'(p) & g_n'(p) \end{pmatrix} \equiv \text{const}, \quad \text{in } [0,1], \tag{49}$$

the bi-orthonormality condition

$$(U_n, V_n) = \frac{1}{\sqrt{2}} \det \begin{pmatrix} \zeta_n(p) & g_n(p) \\ \zeta_n'(p) & g_n'(p) \end{pmatrix} = 1 \tag{50}$$

is satisfied if and only if

$$\zeta_n(\frac{1}{2},p)g_n'(\frac{1}{2},p) - \zeta_n'(\frac{1}{2},p)g_n(\frac{1}{2},p) = \sqrt{2}. \tag{51}$$

The function ζ_n is not uniquely determined by the single condition (51). We must impose a second initial condition at $x = \frac{1}{2}$. Recalling that for n odd the function $g_n(p)$ is even and $g_n'(\frac{1}{2}) = 0$, the function $\zeta_n(p)$ can be chosen such that

$$\zeta_n(\frac{1}{2},p) = 0, \quad \zeta_n'(\frac{1}{2},p) = -\frac{\sqrt{2}}{g_n(\frac{1}{2},p)}. \tag{52}$$

Then, the function $\zeta_n(p)$ is odd with respect to $x = \frac{1}{2}$. Conversely, if n is even, $g_n(p)$ is odd and the function $\zeta_n(p)$ can be chosen such that

$$\zeta_n(\frac{1}{2},p) = \frac{\sqrt{2}}{g_n'(\frac{1}{2},p)}, \quad \zeta_n'(\frac{1}{2},p) = 0, \tag{53}$$

that is $\zeta_n(p)$ is even with respect to $x = \frac{1}{2}$. In conclusion, for n even, $g_n(p)$ is odd and $\zeta_n(p)$ is even, and then the function $(g_n(p)\zeta_n(p))'$ is even. Similarly, when n is odd, $g_n(p)$ is even and $\zeta_n(p)$ is odd, and then the function $(g_n(p)\zeta_n(p))'$ is still even, and the construction of the family $\{V_n\}_{n=0}^{\infty}$ is complete.

3.2 Symmetric Potential and Neumann Boundary Conditions

The method presented in the previous section can be extended to cover the case of Neumann boundary conditions. Consider the eigenvalue problem

$$\begin{cases} y''(x) + \lambda y(x) = q(x)y(x), & \text{in } (0,1), \quad (54)\\ y'(0) = 0 = y'(1), & \quad (55) \end{cases}$$

with $q \in L^2(0,1)$ real-valued. Denote by $\{\lambda_n, g_n\}_{n=0}^{\infty}$ the eigenpairs. Borg (1946) proved the following uniqueness result.

Theorem 3.3 (Borg (1946)). *Let $q \in L^2_{even}(0,1)$. The potential q is uniquely determined by the reduced Neumann spectrum $\{\lambda_n\}_{n=1}^{\infty}$.*

As in the Dirichlet case, the crucial point is to prove that the family $\{1\} \cup \{g_n(q)g_n(p)\}_{n=1}^{\infty}$ is a complete system of functions of $L^2_{even}(0,1)$.

It is worth noticing that the uniqueness result holds without the knowledge of the lower eigenvalue λ_0. Actually, the lower eigenvalue plays a special role in this inverse problem, as it is shown by the following theorem.

Theorem 3.4 (Borg (1946)). *Let $q \in L^2(0,1)$ with $\int_0^1 q = 0$. If the smallest eigenvalue λ_0 of the problem (54)-(55) is zero, then $q \equiv 0$ in $(0,1)$.*

In fact, let us denote by y_0 the eigenfunction associated to the smallest eigenvalue λ_0. By oscillatory properties of Neumann eigenfunctions, y_0 does not vanish in $[0,1]$. Then, we can divide the differential equation (54) by y_0 and integrate by parts in $(0,1)$:

$$0 = \int_0^1 q = \int_0^1 \frac{y_0''}{y_0} = \int_0^1 \left(\frac{y_0'}{y_0}\right)' + \left(\frac{y_0'}{y_0}\right)^2 = \int_0^1 \left(\frac{y_0'}{y_0}\right)^2. \tag{56}$$

By (56) we get $y_0' \equiv 0$ in $[0,1]$ and then $q \equiv 0$ in $(0,1)$.

It should be noted that the above uniqueness results can not be extended, in general, to the Sturm-Liouville problem with even slightly different boundary conditions, for example

$$\begin{cases} \alpha y(0) + y'(0) = 0, & (57) \\ \gamma y(1) + y'(1) = 0, & (58) \end{cases}$$

where α, $\gamma \in \mathbb{R}$. Indeed, if $\alpha \neq 0$ and $\gamma \neq 0$, functions analogous to functions U_n, V_n introduced in the proof of Theorem 3.1 are not necessarily symmetrical with respect to the mid-point $x = \frac{1}{2}$. Borg (1946) gave a counterexample in which the eigenvalue problem with the boundary conditions of the type (57)-(58) does not lead to a complete set of functions in $L^2_{even}(0,1)$.

3.3 Generic L^2 Potential

The uniqueness results addressed in the two preceding sections show that the set of functions $\{g_n(q)g_n(p)\}_{n=0}^{\infty}$, where $g_n(q)$, $g_n(p)$ are the nth eigenfunction corresponding either to Dirichlet or Neumann boundary conditions for potential q and p respectively, are complete in $L^2_{even}(0,1)$, that is in a space of functions which, roughly speaking, has half dimension of the whole space $L^2(0,1)$. To deal with generic $L^2(0,1)$-potentials, the idea by Borg (1946) was to associate to the original Sturm-Liouville problem another Sturm-Liouville problem such that the set of functions $\{g_n(q)g_n(p)\}_{n=0}^{\infty}$ of the original problem together with the set of functions $\{\overline{g}_n(q)\overline{g}_n(p)\}_{n=0}^{\infty}$ of the associated problem form a complete set of $L^2(0,1)$. In particular, the boundary conditions of the associated problem are chosen so that they produce an asymptotic spectral behavior sufficiently different from that of the initial problem.

Let $q \in L^2(0,1)$ be a real-valued potential. As an example, we shall consider the Sturm-Liouville problem

$$\begin{cases} y''(x) + \lambda y(x) = q(x)y(x), & \text{in } (0,1), \quad (59) \\ y(0) = 0, & (60) \\ y(1) = 0 & (61) \end{cases}$$

and its *associate* eigenvalue problem

$$\begin{cases} \overline{y}''(x) + \lambda \overline{y}(x) = q(x)\overline{y}(x), & \text{in } (0,1), \quad (62) \\ \overline{y}(0) = 0, & (63) \\ \gamma \overline{y}(1) + \overline{y}'(1) = 0, & (64) \end{cases}$$

where $\gamma \in \mathbb{R}$. Let us denote by $\{\lambda_n(q), y_n(q) = y_n(x, q, \lambda_n(q))\}_{n=0}^{\infty}$ and by $\{\overline{\lambda}_n(q), \overline{y}_n(q) = \overline{y}_n(x, q, \overline{\lambda}_n(q))\}_{n=0}^{\infty}$ the eigenpairs (with normalized eigenfunctions) of the eigenvalue problems (59)-(61) and (62)-(64), respectively. Let us introduce the two companions Sturm-Liouville problems related respectively to (59)-(61) and (62)-(64) with potential $p \in L^2(0,1)$, namely

$$\begin{cases} z''(x) + \lambda z(x) = p(x)z(x), & \text{in } (0,1), \quad (65) \\ z(0) = 0, & (66) \\ z(1) = 0 & (67) \end{cases}$$

and its *associate* eigenvalue problem

$$\begin{cases} \overline{z}''(x) + \lambda \overline{z}(x) = p(x)\overline{z}(x), & \text{in } (0,1), \quad (68) \\ \overline{z}(0) = 0, & (69) \\ \gamma \overline{z}(1) + \overline{z}'(1) = 0. & (70) \end{cases}$$

Let $\{\lambda_n(p), z_n(p) = z_n(x, p, \lambda_n(p))\}_{n=0}^{\infty}$, $\{\overline{\lambda}_n(p), \overline{z}_n(p) = \overline{z}_n(x, p, \overline{\lambda}_n(p))\}_{n=0}^{\infty}$ be the eigenpairs (with normalized eigenfunctions) of the eigenvalue problems (65)-(67) and (68)-(70), respectively.

Borg (1946) proved the following uniqueness result by two spectra.

Theorem 3.5 (Borg (1946)). *Let q, $p \in L^2(0,1)$. Under the above notation, if $\lambda_n(q) = \lambda_n(p)$ and $\overline{\lambda}_n(p) = \overline{\lambda}_n(q)$ for every $m \geq 0$, then $p = q$.*

The proof follows the same lines of the proof of Theorem 3.1.

4 Uniqueness Towards Stability

4.1 Hochstadt's Formula

In previous section some uniqueness results for the Sturm-Liouville inverse problem have been presented. We have seen that uniqueness for generic L^2 potentials is guaranteed by the knowledge of two complete spectra associated to different sets of boundary conditions. This situation is difficult to meet in practice. Here we shall consider a case in which two potentials share a full spectrum and differ only for a finite number of eigenvalues of a second spectrum. Hochstadt (1973) proved that the difference between two potentials can be written explicitly in terms of certain solutions of the Sturm-Liouville operators. This result suggested an algorithm for solving the inverse problem with finite data and to achieve stability results, see Hochstadt (1977) and Hald (1978).

Consider the Sturm-Liouville problem $(L; \alpha, \beta)$

$$\begin{cases} Lu \equiv -u'' + qu = \lambda u, & \text{in } (0,1), \quad (71) \\ u(0)\cos\alpha + u'(0)\sin\alpha = 0, & (72) \\ u(1)\cos\beta + u'(1)\sin\beta = 0, & (73) \end{cases}$$

with $\alpha, \beta \in \mathbb{R}$ and where $q \in L^1(0,1)$ is a real-valued potential. Denote by $\{\lambda_i\}$ the spectrum of $(L; \alpha, \beta)$.

Replace the boundary condition (73) with the new boundary condition

$$u(1)\cos\gamma + u'(1)\sin\gamma = 0, \tag{74}$$

where $\gamma \in \mathbb{R}$ is such that $\sin(\gamma - \beta) \neq 0$. Denote by $\{\lambda'_i\}$ the spectrum of $(L; \alpha, \gamma)$.

Consider now a second Sturm-Liouville problem $(\widetilde{L}; \alpha, \beta)$ with

$$\widetilde{L}u \equiv -u'' + \widetilde{q}u = \lambda u, \tag{75}$$

with $\widetilde{q} \in L^1(0,1)$. Denote by $\{\widetilde{\lambda}_i\}$ and $\{\widetilde{\lambda}'_i\}$ the spectrum of $(\widetilde{L}; \alpha, \beta)$ and $(\widetilde{L}; \alpha, \gamma)$, respectively.

Assume that

$$\widetilde{\lambda}_i \neq \lambda_i, \quad \mathrm{i} \in \Lambda_0, \ \Lambda_0 = \text{finite set of indices}, \tag{76}$$

$$\widetilde{\lambda}_i = \lambda_i, \quad \mathrm{i} \in \Lambda = \mathbb{N} \setminus \Lambda_0 \tag{77}$$

and

$$\widetilde{\lambda}'_i = \lambda'_i, \quad \text{for every i}. \tag{78}$$

Then

$$q - \widetilde{q} = \sum_{\Lambda_0} (\widetilde{y}_n w_n)', \quad \text{a.e. in } (0,1), \tag{79}$$

where $\widetilde{y}_n$ is a suitable solution of the differential equation

$$\widetilde{y}''_n + (\lambda_n - \widetilde{q})\widetilde{y}_n = 0, \quad \text{in } (0,1), \tag{80}$$

and w_n is the nth eigenfunction of $(L;\alpha,\beta)$. If $\Lambda_0 = \emptyset$, than $q = \widetilde{q}$ a.e. in $(0,1)$ and we obtain the uniqueness for L^1-potentials from two spectra.

Instead of giving the detailed proof of the Hochstadt's result, we will see the main idea of the method and the key points of the analysis. Let us denote by $w = w(x,\lambda)$ the solution of the Cauchy problem

$$\begin{cases} w'' + (\lambda - q)w = 0, \quad \text{in } (0,1), & (81) \\ w(0) = \sin\alpha, & (82) \\ w'(0) = -\cos\alpha, & (83) \end{cases}$$

and by $\widetilde{w} = \widetilde{w}(x,\lambda)$ be the solution of

$$\begin{cases} \widetilde{w}'' + (\lambda - \widetilde{q})\widetilde{w} = 0, \quad \text{in } (0,1), & (84) \\ \widetilde{w}(0) = \sin\alpha, & (85) \\ \widetilde{w}'(0) = -\cos\alpha. & (86) \end{cases}$$

Observe that $w_n = w(x,\lambda_n)$ and $\widetilde{w}_n = \widetilde{w}(x,\widetilde{\lambda}_n)$ are the nth eigenfunction of $(L;\alpha,\beta)$ and $(\widetilde{L};\alpha,\beta)$, respectively. Since we have no information of $\lambda_n \in \Lambda_0$, we try to characterize the relation between the eigenfunctions w_n and $\widetilde{w}_n$ for $n \in \Lambda$. Let us introduce the subspaces of $L^2(0,1)$

$$H = \left\{ f \in L^2(0,1) | \ (f, w_i) = 0, \ i \in \Lambda_0 \right\}, \tag{87}$$

$$\widetilde{H} = \left\{ f \in L^2(0,1) | \ (f, \widetilde{w}_i) = 0, \ i \in \Lambda_0 \right\} \tag{88}$$

and define the linear operator

$$T : H \to \widetilde{H} \tag{89}$$

such that

$$T(w_n) = \widetilde{w}_n, \quad n \in \Lambda. \tag{90}$$

The operator T turns out to be bounded and invertible. The key point of the method is the characterization of the operator T. This can be done in two main steps:

i) By the partial knowledge of the "first" spectrum $\{\lambda_i\}_{i\in\Lambda}$ one can derive the identity

$$(\lambda - \widetilde{L})T(\lambda - L)^{-1} = T, \quad \text{for every } \lambda \in \mathbb{C}, \ \lambda \neq \lambda_i, \ i \in \Lambda. \tag{91}$$

ii) By the full knowledge of the "second" spectrum $\{\lambda_i'\}$, the structure of the operator T can be determined.

Finally, the expression (79) follows by replacing the structure of T found in step ii) in (91).

Concerning step i), to find $(\lambda - T)^{-1}$ we need to solve the equation

$$(\lambda - L)u = f, \quad \lambda \in \mathbb{C} \setminus \{\lambda_n\}_{n \in \mathbb{N}} \tag{92}$$

for every $f \in H$ and under the boundary conditions (α, β). By expressing f by means of its eigenfunction expansion $f = \sum_\Lambda f_n w_n$, $f_n = \frac{(f, w_n)}{\|w_n\|_{L^2}^2}$, on the basis $\{w_n\}$ of $L^2(0,1)$ (with $(w_n, w_k) = \int_0^1 w_n w_k = \|w_n\|_{L^2}^2 \delta_{nk}$), we get

$$u = (\lambda - L)^{-1} f = \sum_\Lambda \frac{f_n}{\lambda - \lambda_n} w_n. \tag{93}$$

Recalling that T is bounded we have

$$T(\lambda - L)^{-1} f = \sum_\Lambda \frac{f_n}{\lambda - \lambda_n} \widetilde{w}_n. \tag{94}$$

The right hand side of (94) belongs to the range of $\widetilde{L}$ and then, recalling that $\lambda_n = \widetilde{\lambda}_n$ for $n \in \Lambda$, we have

$$(\lambda - \widetilde{L}) T (\lambda - L)^{-1} f = \sum_\Lambda \frac{f_n(\lambda \widetilde{w}_n - \widetilde{L}\widetilde{w}_n)}{\lambda - \lambda_n} = \sum_\Lambda f_n \widetilde{w}_n = Tf, \tag{95}$$

for every $f \in H$, which implies the identity (91).

Concerning step ii), we shall introduce an alternative representation of the operator T via the Green function of the problem $(L; \alpha, \beta)$. Let v be the solution of

$$\begin{cases} (\lambda - L)v = v'' + (\lambda - q)v = 0, \quad \text{in } (0,1), & (96) \\ v(1) = -\sin\beta, & (97) \\ v'(1) = \cos\beta. & (98) \end{cases}$$

Then, for $\lambda \neq \lambda_n$, the solution $u = (\lambda - L)^{-1} f$ of

$$\begin{cases} (\lambda - L)u = f, \quad \text{in } (0,1), & (99) \\ u(0)\cos\alpha + u'(0)\sin\alpha = 0, & (100) \\ u(1)\cos\beta + u'(1)\sin\beta = 0 & (101) \end{cases}$$

can be written in terms of the Green function $G(x, y; \lambda)$ as

$$u(x, \lambda) = \int_0^1 G(x, y; \lambda) f(y) dy, \tag{102}$$

where

$$G(x,y;\lambda) = \frac{w(x_<)v(x_>)}{\omega(\lambda)} = \begin{cases} \frac{w(x)v(y)}{\omega(\lambda)} & x \le y, \\ \frac{w(y)v(x)}{\omega(\lambda)} & x \ge y, \end{cases} \tag{103}$$

$$x_< = \min(x,y), \quad x_> = \max(x,y), \tag{104}$$

$$\omega(\lambda) = w(1,\lambda)\cos\beta + w'(1,\lambda)\sin\beta \tag{105}$$

and the function w has been defined in (81)-(83). The function $\omega = \omega(\lambda)$ is an entire function of order $\frac{1}{2}$ in λ and its zeros are the eigenvalues of $(L;\alpha,\beta)$. Since $w_n = w(x,\lambda_n)$ and $v_n = v(x,\lambda_n)$ are eigenfunctions associated to the (simple) eigenvalue λ_n of $(L;\alpha,\beta)$, we have

$$k_n w_n(x) = v_n(x), \tag{106}$$

where k_n is a no vanishing constant given by

$$k_n = -\sin\beta / w_n(1) \quad \text{if } \sin\beta \neq 0, \quad k_n = 1/w_n'(1) \quad \text{if } \sin\beta = 0. \tag{107}$$

Note that $\|w_n\|^2_{L^2} = \frac{\omega'(\lambda_n)}{k_n}$.

By the asymptotic behavior of $w(x,\lambda)$, $v(x,\lambda)$ and $\omega(\lambda)$ as $\lambda \to \infty$, it can be shown that the Green function $G(x,y;\lambda)$ can be expressed by the Mittlag-Leffler expansion

$$G(x,y;\lambda) = \sum_{n=0}^{\infty} \frac{w_n(x_<)v_n(x_>)}{\omega'(\lambda_n)(\lambda-\lambda_n)}, \tag{108}$$

see Ahlfors (1984). Then, for every $f \in H$ we have

$$u(x,\lambda) = (\lambda - L)^{-1} f = \sum_{\Lambda}^{\infty} \frac{k_n w_n(x) \int_0^1 w_n(y) f(y) dy}{\omega'(\lambda_n)(\lambda-\lambda_n)}, \tag{109}$$

and applying T we get

$$T(\lambda - L)^{-1} f = \sum_{\Lambda} \frac{k_n \widetilde{w}_n(x) \int_0^x f(y) w_n(y) dy + \widetilde{w}_n(x) \int_x^1 f(y) v_n(y) dy}{\omega'(\lambda_n)(\lambda-\lambda_n)}. \tag{110}$$

Let $\widetilde{w}_n$, $\widetilde{v}_n$ be the solutions of the initial value problems (81)-(83), (96)-(98), respectively, with q replaced by $\widetilde{q}$ and λ_n replaced by $\widetilde{\lambda}_n$. Then, as in (106), there exists $\widetilde{k}_n \neq 0$ such that $\widetilde{k}_n \widetilde{w}_n(x) = \widetilde{v}_n(x)$.

Exactly on this point, we make use of the second (full) spectrum (e.g., $\lambda_i' = \widetilde{\lambda}_i'$ for every i) to proof the crucial fact

$$\widetilde{k}_n = k_n, \quad \text{for every } n \in \Lambda. \tag{111}$$

To prove this identity, let us introduce the frequency function

$$\nu(\lambda) = w(1)\cos\gamma + w'(1)\sin\gamma \tag{112}$$

for $(L;\alpha,\gamma)$. The function $\nu = \nu(\lambda)$ is an entire function of order $\frac{1}{2}$ in λ and its zeros are the eigenvalues of the problem $(L;\alpha,\gamma)$. By a well-known property of the entire functions of order $\frac{1}{2}$ (see Ahlfors (1984), p. 208) we have

$$\nu(\lambda) = a\prod_{i=0}^{\infty}\left(1 - \frac{\lambda}{\lambda_i'}\right), \tag{113}$$

where a is a suitable constant. (Without loss of generality, we have assumed $\lambda_i' \neq 0$ for every i.) Similarly, we can define the function $\widetilde{\nu}(\lambda)$ for operator $\widetilde{L}$ under boundary conditions and (α,γ), respectively. We have

$$\widetilde{\nu}(\lambda) = \widetilde{a}\prod_{i=0}^{\infty}\left(1 - \frac{\lambda}{\widetilde{\lambda}_i'}\right), \tag{114}$$

where $\widetilde{a}$ is a suitable constant. But, by assumption, $\lambda_i' = \widetilde{\lambda}_i'$ for every i and then, also recalling the asymptotic estimates of eigenvalues, we have

$$\nu(\lambda) = \widetilde{\nu}(\lambda). \tag{115}$$

Now, by evaluating $\omega(\lambda)$ and $\nu(\lambda)$ at $\lambda = \lambda_n$, we have

$$\begin{cases} w_n(1)\cos\beta + w_n'(1)\sin\beta = 0, & (116)\\ w_n(1)\cos\gamma + w_n'(1)\sin\gamma = \nu(\lambda_n). & (117)\end{cases}$$

Solving the linear system with respect to $w_n(1)$ and recalling (107) we get

$$k_n = \frac{\sin(\gamma-\beta)}{\nu(\lambda_n)}. \tag{118}$$

Using the same procedure for the operator $\widetilde{L}$ we obtain

$$\widetilde{k}_n = \frac{\sin(\gamma-\beta)}{\widetilde{\nu}(\lambda_n)} \tag{119}$$

and recalling (115) we finally have $k_n = \widetilde{k}_n$ for every $n \in \Lambda$.

We are now in position to conclude the proof of Hochstadt's formula (79). By substituting (111) in (110) we get

$$T(\lambda - L)^{-1}f = \sum_{\Lambda}\frac{\widetilde{v}_n(x)\int_0^x f(y)w_n(y)dy + \widetilde{w}_n(x)\int_x^1 f(y)v_n(y)dy}{\omega'(\lambda_n)(\lambda-\lambda_n)}. \tag{120}$$

The subtle point is now to realize that the right hand side of (120) is closely related to the Mittag-Leffler expansion of the function

$$g(x)=\frac{\widetilde{v}(x)\int_0^x f(y)w(y)dy+\widetilde{w}(x)\int_x^1 f(y)v(y)dy}{\omega(\lambda)}, \tag{121}$$

where $\widetilde{v}(x)$ is solution of (96)-(98) with q replaced by $\widetilde{q}$, and $\widetilde{w}(x)$ is solution of (84)-(86). In fact, we can represent $g(x)$ in Mittag-Leffler series distinguishing between residues related to poles belonging to Λ_0 and to poles belonging to Λ:

$$\begin{aligned} g(x)=\sum_{\Lambda_0}&\frac{\widetilde{u}_n(x)\int_0^x f(y)w_n(y)dy+\widetilde{z}_n(x)\int_x^1 f(y)v_n(y)dy}{\omega'(\lambda_n)(\lambda-\lambda_n)}+\\ &+\sum_{\Lambda}\frac{\widetilde{v}_n(x)\int_0^x f(y)w_n(y)dy+\widetilde{w}_n(x)\int_x^1 f(y)v_n(y)dy}{\omega'(\lambda_n)(\lambda-\lambda_n)}, \end{aligned} \tag{122}$$

where, to simplify the notation, we define

$$\widetilde{u}_n=\widetilde{v}(x,\lambda_n), \quad \widetilde{z}_n=\widetilde{w}(x,\lambda_n), \quad n\in\Lambda_0. \tag{123}$$

The second term on the right hand side of (121) coincides with an expression present on the right hand side of (120). Now, we replace the above equality inside the identity (91) obtaining

$$(\lambda-\widetilde{L})\left\{g(x)-\sum_{\Lambda_0}\frac{\widetilde{u}_n(x)\int_0^x f(y)w_n(y)dy+\widetilde{z}_n(x)\int_x^1 f(y)v_n(y)dy}{\omega'(\lambda_n)(\lambda-\lambda_n)}\right\}=Tf, \tag{124}$$

for every $f\in H$. After a simple but lengthly calculation and recalling the asymptotic estimates of the function involved as $\lambda\to\infty$, we obtain the characterization of the operator T

$$Tf=f-\frac{1}{2}\sum_{\Lambda_0}\widetilde{y}_n(x)\int_0^x w_n(y)f(y)dy, \tag{125}$$

where

$$\frac{1}{2}\widetilde{y}_n(x)=\frac{\widetilde{u}_n(x)-k_n\widetilde{z}_n(x)}{\omega'(\lambda_n)}. \tag{126}$$

The thesis (79) follows by using the representation (125) in (91).

The above result has been extended by Hochstadt (1973) to symmetric operators, that is even potentials and boundary conditions such that

$\beta = -\alpha$. The proof follows the lines we had sketched above. In particular, information on a second spectrum is replaced by the symmetry of the problem.

Extensions to Sturm-Liouville operators of the form

$$Lu = -\frac{1}{r}((pu')' - qu), \tag{127}$$

with suitable a priori assumptions on r, p and q, are also available Hochstadt (1975). In this case, it is quite clear that there is no hope to determine all the coefficients r, p and q by the knowledge of two full spectra (associated to suitable boundary conditions). Hochstadt was able to prove that, under suitable assumptions, the uniqueness theorems still hold modulo a Liouville transformation. The corresponding problem for a taut string has been considered in Hochstadt (1976).

4.2 A Local Stability Result

In the preceding section it has been shown that the potentials associated with two Sturm-Liouville operators which have a spectrum in common and differ in a finite number of eigenvalues of a second spectrum are intimately linked by explicit expression (79). This result raises the question of the well posedness of the inverse problem. The question is to determine, for example, if a "small" perturbation of the finite set of eigenvalues Λ_0 (which are not in common between the two operators) leads to a "small" perturbation of the unknown potential. A first affirmative answer to this question was given by Hochstadt (1977).

Theorem 4.1 (Hochstadt (1977)). *Let $\epsilon > 0$. Under the assumptions stated at the beginning of section 4.1 and by using the same notation, if*

$$|\lambda_i - \widetilde{\lambda}_i| < \epsilon, \quad i \in \Lambda_0, \tag{128}$$

then

$$|q(x) - \widetilde{q}(x)| < K\epsilon, \quad a.e. \ in \ (0,1), \tag{129}$$

where $K > 0$ is a constant depending only on the data of the problem.

By (79) we have

$$q - \widetilde{q} = \sum_{\Lambda_0} (\widetilde{y}_i w_i)', \quad \text{a.e. in } (0,1), \tag{130}$$

where

$$\begin{cases} w_i'' + (\lambda_i - q(x))w_i = 0, \quad \text{in } (0,1), & (131) \\ w_i(0) = \sin\alpha, & (132) \\ w_i'(0) = -\cos\alpha & (133) \end{cases}$$

and

$$\widetilde{y}_i(x) = 2\frac{\widetilde{u}_i(x) - k_i\widetilde{z}_i(x)}{\omega'(\lambda_i)}. \tag{134}$$

In (134), the functions $\widetilde{u}_i$ and $\widetilde{z}_i$ are solutions to

$$\eta'' + (\lambda_i - \widetilde{q}(x))\eta = 0, \quad \text{in } (0,1), \tag{135}$$

with initial conditions respectively given by

$$\widetilde{z}_i(0) = \sin\alpha, \quad \widetilde{z}_i'(0) = -\cos\alpha, \tag{136}$$

$$\widetilde{u}_i(1) = -\sin\beta, \quad \widetilde{u}_i'(1) = \cos\beta. \tag{137}$$

The function $\omega = \omega(\lambda)$ is defined as

$$\omega(\lambda) = w(1,\lambda)\cos\beta + w'(1,\lambda)\sin\beta, \tag{138}$$

where $w(x,\lambda)$ is a solution of (131) with λ_i replaced by λ. Recall that $\omega = \omega(\lambda)$ is an entire function of λ of order $\frac{1}{2}$ and that its zeros are the eigenvalues $\{\lambda_i\}$ of L under the boundary conditions (α,β).

The constant k_i is given by

$$k_i = \frac{\sin(\gamma-\beta)}{\nu(\lambda_i)}, \tag{139}$$

where

$$\nu(\lambda) = w(1,\lambda)\cos\gamma + w'(1,\lambda)\sin\gamma \tag{140}$$

is an entire function of order 1/2 in λ. The zeros of ν are the eigenvalues $\{\lambda_i'\}$ of L under the boundary conditions (α,γ).

By the assumption (128) we have

$$\lambda_i = \widetilde{\lambda}_i + \epsilon\zeta_i, \quad \text{with } |\zeta_i| < 1, \tag{141}$$

and the differential equation (135) can be written as

$$\eta'' + (\widetilde{\lambda}_i - \widetilde{q}(x))\eta = -\epsilon\zeta_i\eta, \quad \text{in } (0,1), \tag{142}$$

Let us first consider the function $\widetilde{z}_i$, which satisfies the boundary condition (136). If $\zeta_i = 0$, then $\widetilde{z}_i$ is an eigenfunction coinciding with $\widetilde{w}_i$. Therefore, we try to find $\widetilde{z}_i$ in the form

$$\widetilde{z}_i = \widetilde{w}_i + \epsilon\widetilde{r}_i. \tag{143}$$

Replacing this expression in (135)–(136), the perturbation $\widetilde{r}_i$ solves the initial value problem

$$\begin{cases} \widetilde{r}_i'' + (\widetilde{\lambda}_i - \widetilde{q}(x))\widetilde{r}_i = -\zeta_i(\widetilde{w}_i + \epsilon\widetilde{r}_i), \quad \text{in } (0,1), & (144)\\ \widetilde{r}_i(0) = 0, & (145)\\ \widetilde{r}_i'(0) = 0. & (146) \end{cases}$$

Then, by well-known results (see Pöschel and Trubowitz (1987), Chapter 1, Theorem 2) we have

$$\widetilde{r}_i(x) = -\zeta_i \int_0^x (y_1(t)y_2(x) - y_2(t)y_1(x))(\widetilde{w}_i(t) + \epsilon\widetilde{r}_i(t))dt, \tag{147}$$

where $\{y_1, y_2\}$ is a set of fundamental solutions of the differential equation $y'' + (\widetilde{\lambda}_i - \widetilde{q})y = 0$ in $(0,1)$ satisfying the initial conditions $y_1(0) = y_2'(0) = 1$ and $y_1'(0) = y_2(0) = 0$. The function $\widetilde{r}_i(x)$ solves the Volterra equation (147) and it is of class C^2 in $(0,1)$. It should be noticed that, in general, $\widetilde{r}_i(x)$ depends also on ϵ. There is no dependence if equation (147) is linearized by neglecting the term $O(\epsilon)$ in the right hand side of (147).

Similarly, we define

$$\widetilde{u}_i = \widetilde{v}_i + \epsilon\widetilde{s}_i, \tag{148}$$

where $\widetilde{s}_i$ solves

$$\begin{cases} \widetilde{s}_i'' + (\widetilde{\lambda}_i - \widetilde{q}(x))\widetilde{s}_i = -\zeta_i(\widetilde{v}_i + \epsilon\widetilde{s}_i), \quad \text{in } (0,1), & (149)\\ \widetilde{s}_i(1) = 0, & (150)\\ \widetilde{s}_i'(1) = 0. & (151) \end{cases}$$

The functions $\widetilde{v}_i$ and $\widetilde{w}_i$ are eigenfunctions of $\widetilde{L}$ (under the boundary conditions (α, β)) associated to the same eigenvalue $\widetilde{\lambda}_i$. Therefore

$$\widetilde{v}_i(x) = \widetilde{k}_i\widetilde{w}_i \tag{152}$$

with

$$\widetilde{k}_i = \frac{\sin(\gamma - \beta)}{\nu(\widetilde{\lambda}_i)}. \tag{153}$$

Since the two spectra $\{\lambda_i'\}$ and $\{\widetilde{\lambda}_i'\}$ (respectively of L and $\widetilde{L}$ under the boundary conditions (α, γ)) coincide, we have $\nu(\lambda) = \widetilde{\nu}(\lambda)$ and

$$|\nu(\lambda_i) - \nu(\widetilde{\lambda}_i)| < |\nu'(\vartheta_i)||\lambda_i - \widetilde{\lambda}_i| < K\epsilon, \tag{154}$$

where $\vartheta_i \in (\lambda_i, \widetilde{\lambda}_i)$ and K is a suitable constant depending only on $\{\lambda_i'\}$. By (139) we have

$$k_i = \frac{\sin(\gamma - \beta)}{\nu(\lambda_i)} = \frac{\sin(\gamma - \beta)}{\nu(\widetilde{\lambda}_i) + \nu'(\vartheta_i)(\lambda_i - \widetilde{\lambda}_i)} = \widetilde{k}_i + \epsilon l_i, \tag{155}$$

where l_i is a suitable number.

By using the above estimates in (134) and by (152) we obtain

$$\widetilde{y}_i(x) = 2\frac{\widetilde{u}_i(x) - k_i\widetilde{z}_i(x)}{\omega'(\lambda_i)} = 2\frac{(\widetilde{v}_i(x) + \epsilon\widetilde{s}_i(x)) - (\widetilde{k}_i + \epsilon l_i)(\widetilde{w}_i(x) + \epsilon\widetilde{r}_i(x))}{\omega'(\lambda_i)} =$$

$$= \frac{2}{\omega'(\lambda_i)}\left[(\widetilde{v}_i(x) - \widetilde{k}_i\widetilde{w}_i(x)) + \epsilon\widetilde{s}_i(x) - \epsilon\widetilde{k}_i\widetilde{r}_i(x) - \epsilon l_i\widetilde{w}_i(x) - \epsilon^2 l_i\widetilde{r}_i(x)\right] =$$

$$= \frac{2}{\omega'(\widetilde{\lambda}_i) + O(\epsilon)}O(\epsilon) \quad (156)$$

and then

$$|\widetilde{y}_i(x)| < K\epsilon. \quad (157)$$

By using similar arguments we find

$$|\widetilde{y}_i'(x)| < K\epsilon. \quad (158)$$

Finally, by (130) the thesis follows. The above results can be also extended to symmetric potentials.

5 A Local Existence Result

The first existence result for the inverse Sturm-Liouville problem is due to Borg (1946). Consider the differential equation

$$y'' + \lambda y = qy, \quad \text{in } (0,1), \quad (159)$$

with

$$q \in L^2(0,1), \quad \int_0^1 q(x)dx = 0, \quad (160)$$

and the two sets of boundary conditions

$$y(0) = 0 = y(1), \quad \text{(Dirichlet)} \quad (161)$$

$$y(0) = 0, \quad \gamma y(1) + y'(1) = 0, \quad \text{(Mixed)} \quad (162)$$

where γ is a real number.

It is well known that the potential q is uniquely determined by the knowledge of the spectrum $\{\lambda_n\}_{n=0}^\infty$ of (159), (161) and of the spectrum $\{\mu_n\}_{n=0}^\infty$ of (159), (162) (see Theorem 3.5). Suppose to perturb the two spectra $\{\lambda_n\}_{n=0}^\infty$ and $\{\mu_n\}_{n=0}^\infty$, and denote by $\{\lambda_n^*\}_{n=0}^\infty$, $\{\mu_n^*\}_{n=0}^\infty$ the perturbed eigenvalues. We wonder if there exists a Sturm-Liouville operator of the form

$$y'' + \lambda^* y = q^* y, \quad \text{in } (0,1), \quad (163)$$

with

$$q^* \in L^2(0,1), \qquad \int_0^1 q^*(x)dx = 0, \tag{164}$$

such that the spectrum of (163) under the boundary conditions (161) and (162) coincides with the given spectrum $\{\lambda_n^*\}_{n=0}^\infty$, $\{\mu_n^*\}_{n=0}^\infty$, respectively. In other words, we assume that the values of $\{\lambda_n^*\}_{n=0}^\infty$ and $\{\mu_n^*\}_{n=0}^\infty$ are given, and the question is whether there exists a potential $q^* \in L^2(0,1)$, $\int_0^1 q^*(x)dx = 0$, such that these two sequences are the eigenvalues for the boundary conditions (161) and (162), respectively.

Borg (1946) proved that if the sequences of eigenvalues $\{\lambda_n^*\}_{n=0}^\infty$, $\{\mu_n^*\}_{n=0}^\infty$ are a sufficiently small perturbation of the spectra $\{\lambda_n\}_{n=0}^\infty$, $\{\mu_n\}_{n=0}^\infty$ corresponding to a Sturm-Liouville problem with potential q, then there exists a potential q^* with the desired property.

The solution of the existence problem will also indicate a way to construct the potential q^* from q based on a successive approximation method. The result of existence has a local character because the perturbation of the two spectra must be sufficiently small.

Theorem 5.1 (Borg (1946)). *Under the above assumptions, there exists a number $\rho > 0$ such that if*

$$\sum_{n=0}^\infty |\lambda_n - \lambda_n^*|^2 + \sum_{n=0}^\infty |\mu_n - \mu_n^*|^2 \le \rho^2, \tag{165}$$

then the sequences $\{\lambda_n^*\}_{n=0}^\infty$, $\{\mu_n^*\}_{n=0}^\infty$ are the spectrum of the (unique) differential equation

$$y'' + \lambda^* y = q^* y, \quad \textit{in } (0,1), \tag{166}$$

with

$$q^* \in L^2(0,1), \qquad \int_0^1 q^*(x)dx = 0. \tag{167}$$

Proof. Let us denote by $z = z(x)$, $z^* = z^*(x)$ the solutions of (159) (potential q), (166) (potential q^*), respectively, and let us order the eigenvalues and the corresponding eigenfunctions of the problems (159),(161) and (159),(162) as follows: the even index $\nu = 2n + 2$ corresponds to the nth eigenpair of (159),(161), while the odd index $\nu = 2n + 1$ is for the nth eigenpair of (159),(162). Therefore, the eigenvalues $\{\lambda_n\}_{n=0}^\infty$, $\{\mu_n\}_{n=0}^\infty$ are collected in a single sequence $\{\lambda_\nu\}_{\nu=1}^\infty$ and the corresponding eigenfunctions are denoted by $\{z_\nu\}_{\nu=1}^\infty$. The eigenfunctions and eigenvalues of the problem

(166) will be ordered similarly, and will be denoted by $\{z_\nu^*\}_{\nu=1}^\infty$ and $\{\lambda_\nu^*\}_{\nu=1}^\infty$, respectively. Therefore, condition (165) is rewritten as

$$\sum_{\nu=1}^{\infty} |\lambda_\nu - \lambda_\nu^*|^2 \leq \rho^2. \tag{168}$$

Let us suppose for the moment that there exists a differential equation (166), namely that there exists a potential q^*, with the desired properties and let us rewrite (166) in the form of perturbation of the reference equation (159)

$$y'' + (\lambda - q)y = (\lambda - \lambda^* + q^* - q)y, \quad \text{in } (0,1), \tag{169}$$

where the perturbation is represented by the right hand side of (169). Note that the eigenfunction z_ν^* is a solution of (169) for $\lambda = \lambda_\nu$ and $\lambda^* = \lambda_\nu^*$, $\nu = 1, 2, ...$

Let ζ_ν be a solution of (159) with $\lambda = \lambda_\nu$ linearly independent of z_ν (= νth eigenfunction of (159) under boundary conditions (161) or (162)) chosen so that

$$D_\nu = \sqrt{\lambda_\nu} \ \text{ if } \lambda_\nu \geq 1, \quad D_\nu = 1 \ \text{ if } \lambda_\nu < 1, \tag{170}$$

where

$$D_\nu = \det \begin{pmatrix} z_\nu(x) & \zeta_\nu(x) \\ z_\nu'(x) & \zeta_\nu'(x) \end{pmatrix}. \tag{171}$$

Define

$$K_\nu(x,t) = z_\nu(x)\zeta_n(t) - z_\nu(t)\zeta_n(x), \tag{172}$$

$$\Phi_\nu(t) = \lambda_\nu - \lambda_\nu^* + q^*(t) - q(t). \tag{173}$$

By (169), the function z_ν^* (=νth eigenfunction of (166) under boundary conditions (161) or (162)) solves the Volterra equation

$$z_\nu^*(x) = z_\nu(x) + \int_0^x \frac{K_\nu(x,t)}{D_\nu} \Phi_\nu(t) z_\nu^*(t) dt. \tag{174}$$

By solving (174) with the method of the successive approximation, z_ν^* can be represented in the form

$$z_\nu^*(x) = z_\nu(x) + \sum_{p=1}^{\infty} U_\nu^{(p)}(x), \tag{175}$$

where

$$U_\nu^{(p)}(x) = K_\nu \diamond (K_\nu \diamond (... \diamond K_\nu(z_\nu))) \equiv K_\nu^p \diamond z_\nu =$$

$$= \int_0^x \frac{1}{D_\nu^p} K(x,x_1) \int_0^{x_1} K(x_1,x_2) ... \int_0^{x_{p-1}} ... z_\nu(x_p) \prod_{q=1}^{p} \Phi_\nu(x_q) dx_q \tag{176}$$

and then

$$z_\nu^*(x) = z_\nu(x)+$$
$$+\sum_{p=1}^{\infty}\frac{1}{D_\nu^p}\int_0^x\int_0^{x_1}...\int_0^{x_{p-1}}K_\nu(x,x_1)\cdot...\cdot K_\nu(x_{p-1},x_p)z_\nu(x_p)\prod_{q=1}^{p}\Phi_\nu(x_q)dx_q. \tag{177}$$

A direct calculation shows that

$$\int_0^1 z_\nu^* z_\nu \Phi_\nu dx = 0. \tag{178}$$

By (177) and (178), for every $\nu = 1, 2, ...$, we get

$$\int_0^1 z_\nu^2(x)\Delta(x)dx + \delta_\nu \int_0^1 z_\nu^2(x)dx+$$
$$+\sum_{p=1}^{\infty}\frac{1}{D_\nu^p}\int_0^1\int_0^{x_1}...\int_0^{x_p}Q_\nu(x_1,...,x_{p+1})\prod_{q=1}^{p+1}\Phi_\nu(x_q)dx_q = 0, \tag{179}$$

where

$$\Delta(x) = q^*(x) - q(x), \tag{180}$$

$$\delta_\nu = \lambda_\nu - \lambda_\nu^*, \tag{181}$$

$$Q_\nu = z_\nu(x_1)K(x_1,x_2)\cdot ... \cdot K_\nu(x_p,x_{p+1})z_\nu(x_{p+1}). \tag{182}$$

Now, the theorem is proved if we are able to prove that there exists a function $q^* \in L^2(0,1)$ that satisfies all the equations (179) and such that $\int_0^1 q^*(x)dx = 0$. Note that the nonlinearity of the equations (179) is due to the terms Φ_ν.

The function q^* can be determined by the successive approximation method. Suppose that the third term of (179) is zero (that is, we start assuming $\Delta_0 \equiv 0$). Then, we need to find the function

$$\Delta_1(x) = q_1^*(x) - q(x) \tag{183}$$

which satisfies

$$\int_0^1 z_\nu^2(x)\Delta_1(x)dx = -\delta_\nu, \quad \nu = 1, 2, ..., \tag{184}$$

and

$$\int_0^1 \Delta_1(x)dx = 0, \quad \nu = 1, 2, ... \tag{185}$$

Here, the eigenfunctions z_ν are normalized such that

$$\int_0^1 z_\nu^2(x)dx = 1, \quad \nu = 1, 2, \dots . \tag{186}$$

Now, we recall that the sequence of functions $\{z_\nu^2\}_{\nu=1}^\infty$ is a complete system of functions in $L^2(0,1)$. We need also the following result.

Theorem 5.2 (Borg (1946)). *Let $\{c_\nu\}_{\nu=1}^\infty$ be any subsequence of real numbers such that*

$$\sum_{\nu=1}^\infty c_\nu^2 < \infty. \tag{187}$$

Let $\{z_\nu^2\}_{\nu=1}^\infty$ the above sequence of functions. There exists a unique function $f \in L^2(0,1)$ such that

$$\int_0^1 f(x)dx = 0 \tag{188}$$

and

$$c_\nu = \int_0^1 f(x) z_\nu^2(x)dx, \quad \nu = 1, 2, \dots . \tag{189}$$

Moreover, there exist two constants m and M independent of f, $0 < m < M < \infty$, such that

$$m^2 \sum_{\nu=1}^\infty c_\nu^2 \le \int_0^1 f^2(x)dx \le M^2 \sum_{\nu=1}^\infty c_\nu^2. \tag{190}$$

By the assumption (168), the condition (187) is satisfied (with $c_\nu = -\delta_\nu$)

$$\sum_{\nu=1}^\infty \delta_\nu^2 = \sum_{\nu=1}^\infty |\lambda_\nu - \lambda_\nu^*|^2 \le \rho^2 < \infty \tag{191}$$

and then, by Theorem 5.2, there exists a unique function $\Delta_1 \in L^2(0,1)$ solution of (184), (185).

Next approximation, $\Delta_2(x)$, can be determined similarly. In general, the function $\Delta_i(x)$ solves for $\nu = 1, 2, \dots$ the equations

$$\int_0^1 z_\nu^2(x)\Delta_i(x)dx + \delta_\nu \int_0^1 z_\nu^2(x)dx +$$
$$+ \sum_{p=1}^\infty \frac{1}{D_\nu^p} \int_0^1 \int_0^{x_1} \dots \int_0^{x_p} Q_\nu(x_1, \dots, x_{p+1}) \prod_{q=1}^{p+1} \Phi_{\nu,i-1}(x_q)dx_q = 0, \tag{192}$$

where

$$\int_0^1 \Delta_i(x) = 0, \tag{193}$$

$$\Phi_{\nu,i-1}(x) = \delta_\nu + \Delta_{i-1}(x), \tag{194}$$

$$q_i^*(x) = q(x) - \Delta_i(x), \quad \Delta_i(x) = q_i^*(x) - q(x), \tag{195}$$

$$\Delta_i \in L^2(0,1). \tag{196}$$

In order to prove the convergence of the iterative procedure we need to show that the sequence $\{\Delta_i\}_{i=0}^\infty$ converges in $L^2(0,1)$. In the following we recall the main steps of the proof of convergence, see Borg (1946) for details. The proof is essentially in three main steps. First of all, one can prove that the sequence $\{\Delta_i\}_{i=0}^\infty$ is bounded in $L^2(0,1)$ for every choice of the eigenvalues $\{\lambda_\nu^*\}_{\nu=1}^\infty$ provided that ρ is sufficiently small. More precisely, for every i there exists a constant $K > 0$ independent of i such that

$$\int_0^1 \Delta_i^2(x)dx \leq K. \tag{197}$$

This condition is satisfied if

$$\rho \leq \frac{1}{8M\sqrt{1+\eta^2}}, \tag{198}$$

where

$$\eta^2 = 16 s_0^2 M^2 M_1^2 (e^{M_1^2} - 1), \tag{199}$$

$$s_0 = \left(\sum_{\nu=1}^\infty \frac{1}{D_\nu^2}\right)^{\frac{1}{2}} \quad (D_\nu = O(\nu),\ D_\nu \geq 1), \tag{200}$$

$$\max_{\nu \geq 1} |K_\nu(x_q, x_{q+1})| \leq M_1, \quad \max_{\nu \geq 1} |z_\nu(x_1) z_\nu(x_{p+1})| \leq M_1, \tag{201}$$

and M is the constant appearing in (190).

The second step consists in proving that the sequence $\{\Delta_i\}_{i=0}^\infty$ is a Cauchy sequence in $L^2(0,1)$. Then, we have

$$\lim_{i\to\infty} \Delta_i(x) = \Delta(x) \in L^2(0,1), \quad \int_0^1 \Delta(x)dx = 0. \tag{202}$$

In the third step it is shown that the function q^* found by the iterative process, e.g. $q^*(x) = q(x) - \Delta(x)$, actually solves the differential equation (166). Finally, by the uniqueness Theorem 3.5, the function q^* is unique and the proof is complete. □

Remark 5.3. Similar result also applies in the case of symmetric potential and when we know only one spectrum for given boundary conditions. Consider, for example, the differential equation $y'' + (\lambda - q(x))y = 0$ in $(0,1)$, with $q \in L^2_{even}(0,1)$ and with spectrum $\{\lambda_n\}_{n=1}^{\infty}$ under Dirichlet boundary conditions $y(0) = 0 = y(1)$. It is possible to determine a differential equation $y'' + (\lambda^* - q^*(x))y = 0$ in $(0,1)$, with $q^* \in L^2_{even}(0,1)$ such that its full Dirichlet spectrum coincides with a given sequence of real numbers $\{\lambda_n^*\}_{n=1}^{\infty}$ provided that

$$\sum_{n=1}^{\infty} |\lambda_n - \lambda_n^*|^2 \leq \rho^2, \tag{203}$$

with ρ "sufficiently" small. The same argument can be also extended to Neumann boundary conditions $y'(0) = 0 = y'(1)$ under prescribed reduced spectra.

Remark 5.4. Generally speaking, Theorem (5.1) is the analogous of the Riesz-Fischer Theorem in the reconstruction of a function on the basis of its Fourier coefficients on a given basis of functions. However, the existence results for Sturm-Liouville operators based on eigenvalue information are valid only locally, namely if the distance ρ is small enough. This limitation seems not easy to remove and probably is due to the need to satisfy the inequality $\lambda_n < \lambda_{n+1}$ for every n, which have not correspondence among the Fourier coefficient representations. An attempt to use the local existence result to prove a global existence of the potential is sketched in Borg (1946).

6 An Euler-Bernoulli Inverse Eigenvalue Problem

Consider a thin straight uniform beam, of length L, resting on an elastic foundation of stiffness $q = q(x)$ (per unit length), $q(x) \geq 0$. Let the beam be made by homogeneous, isotropic linear elastic material with constant Young's modulus $E > 0$ and constant linear mass density $\rho > 0$. The free vibrations of the beam are governed by the partial differential equation

$$EI\frac{\partial^4 v(x,t)}{\partial x^4} + q(x)v(x,t) + \rho\frac{\partial^2 v(x,t)}{\partial^2 t} = 0, \quad (x,t) \in (0,L)\times(0,\infty), \tag{204}$$

where $I > 0$ is the moment of inertia of the cross-section and $v = v(x,t)$ is the transversal displacement of the beam axis at the point of abscissa x evaluated at time t. The beam is supposed to be simply supported at the ends, namely

$$v = 0, \quad \frac{\partial^2 v}{\partial x^2} = 0, \quad \text{at } x = 0 \text{ and } x = L, \text{ for every } t \geq 0. \tag{205}$$

Let us assume that the beam vibrates with frequency ω, i.e. $v(x,t) = u(x)\cos(\omega t)$. Then, the spatial free vibrations are governed by the boundary value problem

$$\begin{cases} Lu \equiv u^{IV} + qu = \lambda u, \quad \text{in } (0,1), & (206) \\ u(0) = 0 = u(1), & (207) \\ u''(0) = 0 = u''(1), & (208) \end{cases}$$

where $\lambda = \omega^2$ and where, to simplify the notation and without loss of generality, we have assumed $L = 1$, $EI = 1$, $\rho = 1$. It is well known that for a given beam and a given *potential* q, q regular enough (e.g., $q \in C([0,1])$, q not necessarily positive), this eigenvalue problem has $\{\lambda_n, g_n(x)\}_{n=1}^{\infty}$ eigenpairs, with simple real eigenvalues such that $\lim_{n\to\infty} \lambda_n = \infty$. For example, if $q \equiv 0$, then $\lambda_n = (n\pi)^4$, $g_n(x) = \sqrt{2}\sin(n\pi x)$ ($\int_0^1 g_n^2(x)dx = 1$), $n \geq 1$.

A reasonable *inverse* eigenvalue problem consists in determining q from the eigenvalues $\{\lambda_n\}_{n=1}^{\infty}$ of the beam. In this section we shall consider a uniqueness problem for symmetric q which is, in some respects, the fourth-order (local) counterpart of the Borg's Theorem 3.1 for Sturm-Liouville operators. The result has been obtained by Elcrat and Papanicolaou (1997) and it is stated in the following theorem. We refer to Gladwell (2004) for a general overview of the inverse problem for the vibrating Euler-Bernoulli beam.

Theorem 6.1 (Elcrat and Papanicolaou (1997)). *Let $q = q(x)$ be a continuous potential in $[0,1]$ such that $q(1-x) = q(x)$ and $\int_0^1 q(x)dx = 0$. There exists a constant Q, $Q > 0$, such that if $\|q\|_\infty = \max_{x\in[0,1]} |q(x)| \leq Q$, then the full spectrum $\{\lambda_n\}_{n=1}^{\infty}$ determines uniquely q.*

The proof is by contradiction. Consider the eigenvalue problem

$$\begin{cases} \widetilde{L}u \equiv u^{IV} + \widetilde{q}u = \lambda u, \quad \text{in } (0,1), & (209) \\ u(0) = 0 = u(1), & (210) \\ u''(0) = 0 = u''(1), & (211) \end{cases}$$

with $\widetilde{q} \in C([0,1])$, $\int_0^1 q = 0$, $q(1-x) = q(x)$ in $[0,1]$, and let $\{\widetilde{\lambda}_n\}_{n=1}^{\infty}$ be the spectrum of $\widetilde{L}$. We shall prove that, under the hypotheses of the theorem, if $\lambda_n = \widetilde{\lambda}_n$ for every $n \geq 1$, then $q = \widetilde{q}$ in $[0,1]$.

Denote by g_n, $\widetilde{g}_n$ the nth eigenfunction of L, $\widetilde{L}$, respectively, with $\int_0^1 g_n^2 dx = \int_0^1 \widetilde{g}_n^2 dx = 1$, $n \geq 1$. By multiplying by $\widetilde{g}_n$ (respectively, by g_n) the differential equation satisfied by g_n (respectively, by $\widetilde{g}_n$), integrating by parts and recalling that $\lambda_n = \widetilde{\lambda}_n$ for every $n \geq 1$, we have

$$\int_0^1 (q - \widetilde{q}) g_n \widetilde{g}_n dx = 0, \quad \text{for every } n \geq 1. \tag{212}$$

Therefore, it is enough to show that $\{g_n\widetilde{g}_n\}_{n=1}^{\infty}$ is a complete family of functions in $L^2_{even}(0,1)$.

Following Borg's approach, Elcrat and Papanicolau introduce the set of functions

$$\begin{cases} U_0(x) = 1, & (213)\\ U_n(x) = \sqrt{2}\left(\int_0^1 g_n\widetilde{g}_n dx - g_n(x)\widetilde{g}_n(x)\right), \quad n \geq 1. & (214)\end{cases}$$

Note that $\{U_n\}_{n=0}^{\infty}$ is a bounded subset of $L^2_{even}(0,1)$, with $\int_0^1 U_n dx = 0$ for every $n \geq 1$. By (212), we have

$$\int_0^1 (q-\widetilde{q})U_n dx = 0, \quad \text{for every } n \geq 0, \tag{215}$$

and therefore the uniqueness result follows if $\{U_n\}_{n=0}^{\infty}$ is a complete family of functions in $L^2_{even}(0,1)$. Actually, it is possible to prove that $\{U_n\}_{n=0}^{\infty}$ is a basis of $L^2_{even}(0,1)$ for $\|q\|_\infty$, $\|\widetilde{q}\|_\infty$ small enough.

Let $\{e_n(x) = \sqrt{2}\cos(2n\pi x)\}_{n=1}^{\infty} \cup \{1\} \equiv \{e_n\}_{n=0}^{\infty}$ be an orthonormal basis of $L^2_{even}(0,1)$. Define the linear operator $A : L^2_{even}(0,1) \to L^2_{even}(0,1)$ as

$$A(e_n) = U_n, \quad n \geq 0. \tag{216}$$

We prove that A is a linear boundedly invertible operator. By general properties, this result follows by showing that A can be written as $A = I+T$, where I is the identity operator and T is a compact operator with $\|T\| < 1$ (see Friedman (1982)).

At this point we need the following asymptotic estimates:

$$|g_n(x) - \sqrt{2}\sin(n\pi x)| \leq C\tfrac{\ln n}{n^3}\|q\|_\infty, \quad \text{as } n \to \infty, \tag{217}$$

$$|\widetilde{g}_n(x) - \sqrt{2}\sin(n\pi x)| \leq \widetilde{C}\tfrac{\ln n}{n^3}\|\widetilde{q}\|_\infty, \quad \text{as } n \to \infty, \tag{218}$$

where C, $\widetilde{C}$ are positive constants depending on Q and bounded as $Q \to 0$, Elcrat and Papanicolaou (1997). Then, by (217), (218) we have

$$|U_n(x) - \sqrt{2}\cos(2n\pi x)| \leq C\frac{\ln n}{n^3}Q, \quad \text{as } n \to \infty, \tag{219}$$

where C is a positive constant depending on Q. We also recall that, if $q \in C^r([0,1])$, $r = 0,1,2,3$, then the eigenvalues λ_n of problem (206)–(208) have the following eigenvalue expansion

$$|\lambda_n - (n\pi)^4| = O\left(\frac{1}{n^r}\right), \text{ as } n \to \infty, \tag{220}$$

see Elcrat and Papanicolaou (1997).

In order to prove that T is compact, it is enough to show that

$$\sum_{n=0}^{\infty} \|Tv_n\|_{L^2}^2 < \infty, \tag{221}$$

for some orthonormal basis $\{v_n\}_{n=1}^{\infty}$ of $L^2_{even}(0,1)$, see Friedman (1982). By choosing $v_n = e_n$, $n \geq 0$, we have

$$\|Te_n\|_{L^2} = O\left(\frac{\ln n}{n^3}\right) \tag{222}$$

and condition (221) is satisfied.

Let $f = \sum_n f_n e_n$ be any element of $L^2_{even}(0,1)$, where $f_n = (f, e_n)$. By definition, the image of f under the operators A, T is given by $Af = \sum_k \left(\sum_n A_{kn} f_n\right) e_k$, $Tf = \sum_k \left(\sum_n T_{kn} f_n\right) e_k$, respectively, where $A_{kn} = (U_n, e_k)$ and $T_{kn} = A_{kn} - \delta_{kn}$. In order to prove that $\|T\|_{L^2}^2 < 1$, we first observe that

$$\|T\|_{L^2}^2 \leq \sum_{k,n}^{\infty} T_{kn}^2. \tag{223}$$

This inequality follows easily by the definition of the L^2-norm of T and by Schwarz's inequality. Now, let us estimate the series on the right end side of (223). We have

$$\sum_{k,n}^{\infty} T_{kn}^2 = \sum_n \left(\sum_k A_{kn}^2 - 1 - 2(A_{nn} - 1)\right). \tag{224}$$

By Parseval's inequality and by using the asymptotic estimate (219), we have

$$|\sum_k^{\infty} A_{kn}^2 - 1| \leq C\frac{1+\ln n}{n^3} Q, \tag{225}$$

where C is a positive constant depending on Q. Moreover, again by (219) we have

$$|A_{nn} - 1| \leq O\left(\frac{\ln n}{n^3}\right) Q, \quad \text{as } n \to \infty. \tag{226}$$

Now, by inserting (225) and (226) in (224) we obtain

$$\sum_{k,n}^{\infty} T_{kn}^2 = \sum_n O\left(\frac{1+\ln n}{n^3}\right) Q. \tag{227}$$

The series on the right end side converges and, therefore, the condition $\|T\|_{L^2}^2 < 1$ is satisfied, provided that Q is small enough. Then, the operator A is invertible and the family $\{U_n\}_{n=0}^{\infty}$ is a complete family of functions in $L^2_{even}(0,1)$. This implies the uniqueness result stated in Theorem 6.1.

We conclude this section by recalling an extension of the above uniqueness result for near-constant fourth-order operator with two unknown coefficients due to Schueller (2001). Consider the eigenvalue problem

$$\begin{cases} u^{IV} - (p_1 u')' + p_2 u = \lambda u, \quad \text{in } (0,1), & (228) \\ u(0) = 0 = u(1), & (229) \\ u''(0) = 0 = u''(1), & (230) \end{cases}$$

where p_1, $p_2 \in L^2(0,1)$. Let

$$\mathcal{P} = \{(p_1, p_2) \in (L^2(0,1))^2 \mid (228) - (230) \text{ has only simple eigenvalues}\}, \tag{231}$$

$$Sym = \{(p_1, p_2) \in \mathcal{P} \mid p_i(x) = p_i(1-x) \text{ a.e. in } (0,1)\} \tag{232}$$

and for any $(a_1, a_2) \in Sym$ let us define

$$\mathcal{B}(a_1, a_2; \epsilon) = \{(p_1, p_2) \in (L^2(0,1))^2 \mid \|p_1 - p_2\|_{L^2} + \|p_2 - a_2\|_{L^2} < \epsilon\}. \tag{233}$$

Theorem 6.2 (Schueller (2001)). *Let $(a_1, a_2) \in Sym$ with a_1, a_2 constants. There exists $\epsilon > 0$ such that if $(p_1, p_2) \in \mathcal{B}(a_1, a_2; \epsilon)$, $(\widetilde{p}_1, \widetilde{p}_2) \in \mathcal{B}(a_1, a_2; \epsilon)$, with $(p_1, p_2) \in Sym$ and $(\widetilde{p}_1, \widetilde{p}_2) \in Sym$, have the same spectrum, then:*

i) $p_2 = \widetilde{p}_2$ implies $p_1 = \widetilde{p}_1$;
ii) $p_1 = \widetilde{p}_1$ and $\int_0^1 (p_2 - \widetilde{p}_2) = 0$ implies $p_2 - \widetilde{p}_2 = 0$.

Extensions of the above results to other sets of boundary conditions (e.g., cantilever beam) or to multi-span beams are open problems.

Bibliography

L.V. Ahlfors. *Complex Analysis.* McGraw-Hill, 1984.

G. Borg. Eine Umkehrung der Sturm-Liouvilleschen Eigenwertaufgabe. Bestimmung der Differentialgleichung durch die Eigenwerte. *Acta Mathematica*, 78:1–96, 1946.

H. Brezis. *Analisi funzionale.* Liguori Editore, 1986.

C.F. Coleman and J.R. McLaughlin. Solution of the inverse spectral problem for an impedance with integrable derivative. Part II. *Communications on Pure and Applied Mathematics*, 46:185–212, 1993.

A. Elcrat and V.G. Papanicolaou. On the inverse problem of a fourth-order self-adjoint binomial operator. *SIAM Journal of Mathematical Analysis*, 28:886–996, 1997.

A. Friedman. *Foundations of Modern Analysis.* Dover, 1982.

I.M. Gel'fand and B.M. Levitan. On the determination of a differential equation from its spectral function. *American Mathematical Society Translation Series 2*, 1:253–304, 1955.

G.M.L. Gladwell. *Inverse Problems in Vibration.* Kluwer Academic Publishers, 2004.

G.M.L. Gladwell and A. Morassi. On isospectral rods, horns and strings. *Inverse Problems*, 11:533–554, 1995.

O.H. Hald. The inverse Sturm-Liouville problem with symmetric potentials. *Acta Mathematica*, 141:263–291, 1978.

O.H. Hald and J.R. McLaughlin. Solution of inverse nodal problems. *Inverse Problems*, 5:307–347, 1989.

H. Hochstadt. The inverse Sturm-Liouville problem. *Communications on Pure and Applied Mathematics*, 26:715–729, 1973.

H. Hochstadt. On inverse problems associated with Sturm-Liouville operators. *Journal of Differential Equations*, 17:220–235, 1975.

H. Hochstadt. On the determination of the density of a vibrating string from spectral data. *Journal of Mathematical Analysis and Applications*, 55:673–585, 1976.

H. Hochstadt. On the well-posedness of the inverse Sturm-Liouville problems. *Journal of Differential Equations*, 23:402–413, 1977.

B.M. Levitan. On the determination of a Sturm-Liouville equation by two spectra. *American Mathematical Society Translation Series 2*, 68:1–20, 1968.

B.M. Levitan. *Inverse Sturm-Liouville Problems.* VNU Science Press, 1987.

J. Pöschel and E. Trubowitz. *Inverse Spectral Theory.* Academic Press, 1987.

A. Schueller. Uniqueness for near-constant data in fourth-order inverse eigenvalue problems. *Journal of Mathematical Analysis and Applications*, 258:658–670, 2001.

E.C. Titchmarsh. *The Theory of Functions.* Oxford University Press, 1939.

E.C. Titchmarsh. *Eigenfunction Expansion Associated with Second-order Differential Equations. Part I.* Clarendon Press, 1962.

A Least Squares Functional for Solving Inverse Sturm-Liouville Problems

Norbert Röhrl

Department of Mathematics and Physics, University of Stuttgart, Stuttgart, Germany

Abstract We present a method to numerically solve the Sturm-Liouville inverse problem using least squares following (Röhrl, 2005, 2006). We show its merits by computing potential and boundary conditions from two sequences of spectral data in several examples. Finally we prove theorems which show why this approach works particularly well.

1 Inverse Problems with Least Squares

General inverse problems can be formulated in terms of least squares as follows.

Let $f_{p_1,\dots,p_n} : \mathbb{R} \to \mathbb{R}$ be a function with parameters $\mathbf{p} = p_1, \dots, p_n$. Given a set of measurements $\{(i, x_i, f_i)\}$, find parameters which minimize

$$G(\mathbf{p}) = \sum_i \left(f_{p_1,\dots,p_n}(x_i) - f_i\right)^2 .$$

The function f is a model for the physical system which we are measuring. By minimizing the least squares sum, we choose the optimal set of parameters to reproduce the measurements with minimal error in the l^2-norm. If the function $G(\mathbf{p})$ is sufficiently nice, we can use a gradient minimization scheme to compute parameters $\mathbf{p}_0$ with small least squares error $G(\mathbf{p}_0)$.

Note that in general there is no unique solution and there might even be no satisfactory solution, if the model f is not suitable. In these notes we will prove that this method indeed works for inverse Sturm-Liouville problems. But let us first think about what could go wrong in the general case.

1.1 Gradient Flow

Numerical gradient minimization algorithms approximate the gradient flow $\mathbf{p}(t)$ of G given by the equation

$$\frac{\mathrm{d}\mathbf{p}(t)}{\mathrm{d}t} = -\nabla G(\mathbf{p}(t)),$$

where ∇G denotes the gradient of the least squares sum G. The gradient flow $\mathbf{p}(t)$ is a curve in parameter space, which always points in the direction of steepest descent.

Let us use two one dimensional functions to show the problems we can run into when using gradient based minimization. First, there might be local minima:

Consider the function $G : \mathbb{R} \to \mathbb{R}, x \mapsto \sin\left(\frac{3\pi}{1+x^2}\right)$

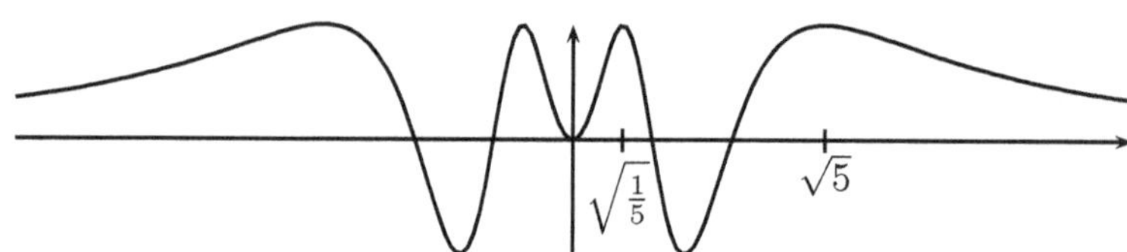

There are three cases when starting the minimization procedure at x_0 and following the gradient:

- $|x_0| < \sqrt{\frac{1}{5}}$ leads to the local minimum at $x = 0$,
- $|x_0| > \sqrt{5}$ will lead to infinity,
- only an x_0 in between will lead to one of the two global minima.

Second, the speed of convergence can be very slow, if the neighborhood of the minimum is flat.

Consider the function $G : \mathbb{R} \to \mathbb{R}, x \mapsto x^n$ with gradient flow

$$\frac{\mathrm{d}\mathbf{p}(t)}{\mathrm{d}t} = -n\mathbf{p}(t)^{n-1}.$$

For $n = 2$ this gives

$$\mathbf{p}(t) = \mathbf{p}_0 \mathrm{e}^{-2t} \text{ and } G(\mathbf{p}(t)) = \mathbf{p}_0^2 \mathrm{e}^{-4t}.$$

For $n > 2$ we get a Bernoulli DE with solution

$$\mathbf{p}(t) = (n(n-2)t + \mathbf{p}_0^{2-n})^{-\frac{1}{n-2}} \text{ and } G(\mathbf{p}(t)) = \mathbf{p}_0^{n(2-n)}(n(n-2)t)^{-\frac{n}{n-2}}.$$

This means convergence to zero in this case could get as slow as t^{-1}.

Now we have laid down the principles of our approach. In the following chapters we will apply these ideas to the inverse Sturm-Liouville problem.

2 The Inverse Sturm-Liouville Problem

We consider the Sturm-Liouville equation

$$-u'' + q(x)u = \lambda u \tag{1}$$

on $[0, 1]$ with $q(x) \in L^2([0, 1], \mathbb{R})$ real, and separated boundary conditions given by two angles α, β

$$u(0)\cos\alpha + u'(0)\sin\alpha = 0, \quad u(1)\cos\beta + u'(1)\sin\beta = 0\,. \tag{2}$$

For each boundary condition $(\alpha\beta)$ there exists an infinite sequence of eigenvalues $\lambda_{q,(\alpha\beta),n}$, $n \in \mathbb{N}$ (Titchmarsh, 1962). The inverse problem is to compute the potential q and the boundary conditions α, β from (partial) spectral data $\lambda_{q,(\alpha\beta),n}$.

Before we think about recovering a potential, we have to clarify existence and uniqueness of a potential which satisfies a given set of spectral data.

We start with uniqueness, which tells us how much spectral information we need to determine the potential. The answer was first given by Borg (Borg, 1946) and generalized by Levinson (Levinson, 1949), and tells us, that we need two sequences of spectra. Fix three angles α, β, and γ with $\sin(\beta - \gamma) \neq 0$. Let $\lambda_{q,i,n}$, $i \in \{1, 2\}$ denote the eigenvalues with respect to the boundary conditions $(\alpha\beta)$ resp. $(\alpha\gamma)$.

Theorem 2.1 ((Borg, 1946; Levinson, 1949)). *Given potentials $Q, q \in L^1[0, 1]$ with*

$$\lambda_{Q,i,n} = \lambda_{q,i,n} \qquad \textit{for all } n \in \mathbb{N}, \quad i = 1, 2\,,$$

then $Q = q$.

The existence question is for which input data there exists a potential with the given spectrum. In other words: for which sequences $(\lambda_{Q,i,n})$ there is a q with $\lambda_{q,i,n} = \lambda_{Q,i,n}$? The answer is affirmative, if the following three conditions are met:

- The two spectra interlace, i.e. $\lambda_{Q,1,n} < \lambda_{Q,2,n} < \lambda_{Q,1,n+1}$ for all $n \in \mathbb{N}$
- Both spectra satisfy the asymptotic formulas.
- The boundary conditions fulfill
  1. $\sin\alpha, \sin\beta, \sin\gamma \neq 0$ (Levitan, 1987) or
  2. $\alpha = \beta = 0, \sin\gamma \neq 0$ (Dahlberg and Trubowitz, 1984)

Now we can formulate the least squares functional for the inverse Sturm-Liouville problem. Given boundary conditions $(\alpha\beta)$, $(\alpha\gamma)$ and partial spectral data $\lambda_{Q,i,n}$ with (i,n) in a finite set $I \subseteq \{1,2\} \times \mathbb{N}$ of an unknown potential Q, we define the functional

$$G(q) = \sum_{(n,i)\in I} (\lambda_{q,i,n} - \lambda_{Q,i,n})^2$$

for a trial potential q. We immediately see that $G(q)$ is non-negative and zero iff $\lambda_{q,i,n} = \lambda_{Q,i,n}$ for all $(n,i) \in I$.

To compute the gradient of G we need the derivative of the eigenvalues wrt. the potential. If $g_{q,i,n}$ denotes the eigenfunction of L^2-norm 1 corresponding to the eigenvalue $\lambda_{q,i,n}$, then the derivative of $\lambda_{q,i,n}$ in direction h is (Pöschel and Trubowitz, 1987)

$$\frac{\partial}{\partial q}\lambda_{q,i,n}[h] = \int_0^1 h g_{q,i,n}^2 \,\mathrm{d}x\,.$$

and therefore

$$\frac{\partial}{\partial q}G[h](q) = 2 \sum_{(n,i)\in I} \int_0^1 (\lambda_{q,i,n} - \lambda_{Q,i,n})\, g_{q,i,n}^2 h \,\mathrm{d}x\,.$$

Hence we have the gradient

$$\nabla G(q) = 2 \sum_{(n,i)\in I} (\lambda_{q,i,n} - \lambda_{Q,i,n})\, g_{q,i,n}^2\,.$$

Later we will prove the following theorem, which immediately implies the absence of local minima.

Theorem 2.2. *The set of squared eigenfunctions*

$$\{g_{i,n}^2 | (i,n) \in \{1,2\} \times \mathbb{N}\}$$

is linearly independent in H^1.

Corollary 2.3. *The functional G has no local minima at q with $G(q) > 0$, i.e.*

$$\frac{\partial}{\partial q}G[h](q) = 0 \;\forall h \Longleftrightarrow \nabla G(q) = 0 \Longleftrightarrow G(q) = 0\,.$$

Proof. Since the $g_{i,n}^2$ are linearly independent, the gradient $\nabla G(q)$ is zero if and only if all differences $\lambda_{q,i,n} - \lambda_{Q,i,n}$ vanish. This is obviously equivalent to $G(q) = 0$. □

Later we also show, that the algorithm converges exponentially when we start with a potential which is sufficiently close to a solution. The precise statement for the convergence of the gradient flow is:

Theorem 2.4. *If $G(q_0)$ is sufficiently small, there exists a constant $d > 0$ such that*

$$G(q_t) \leq G(q_0)\mathrm{e}^{-dt}$$

for all $t \geq 0$.

2.1 Examples

We will now give a number of examples first published in (Röhrl, 2005). The conjugate gradient algorithm can be summarized as follows (see e.g. (Press et al., 1992) for a practical introduction).

1. Start with test potential q_0, $i = 0$.
2. While $G(q_i)$ too big:
    a) Compute eigenvalues and eigenfunctions of q_i.
    b) Compute $G(q_i)$ and $\nabla G(q_i)$.
    c) Choose a direction h_i using conjugate gradients.
    d) Find a minimum along the line $G(q_i + \alpha h_i)$.
    e) $q_{i+1} = q_i + \alpha_0 h_i$

In our examples we always start with $q_0 = 0$, use 30 pairs of eigenvalues, and count one loop of the conjugate gradient algorithm as one iteration. In the figures we use $\Delta_2 = \|Q - q\|_2$ to denote the L^2-distance of our approximation to the correct potential, iter# for the number of iterations, and $\triangle_\lambda = \max_{(i,n)\in I} |\lambda_{q,i,n} - \lambda_{Q,i,n}|$ to measure how far the eigenvalues of our approximation differ from the given ones.

First we use Dirichlet-Dirichlet and Dirichlet-Neumann boundary conditions ($\alpha = \beta = 0$, $\gamma = \pi/2$). The graph below is the result after 5 iterations. Further iterations do not change the graph visibly.

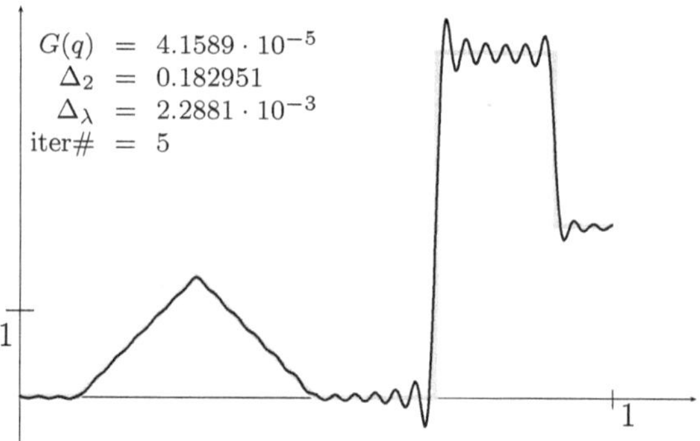

Below we plot the values of $G(q)$ and the difference $\Delta_2 = \|q - Q\|_2$ of the L^2-norms of our approximation and the true potential. The left plot is logarithmic and we can see the exponential convergence in the number of iterations.

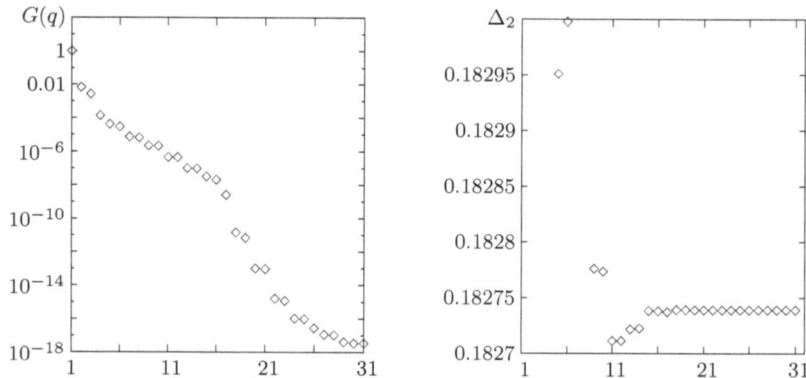

Changing the boundary conditions to $\alpha = \beta = \pi/4$, $\gamma = -\pi/4$ does influence the speed of convergence significantly, as the following example shows. Except for the boundary conditions this everything is exactly as before.

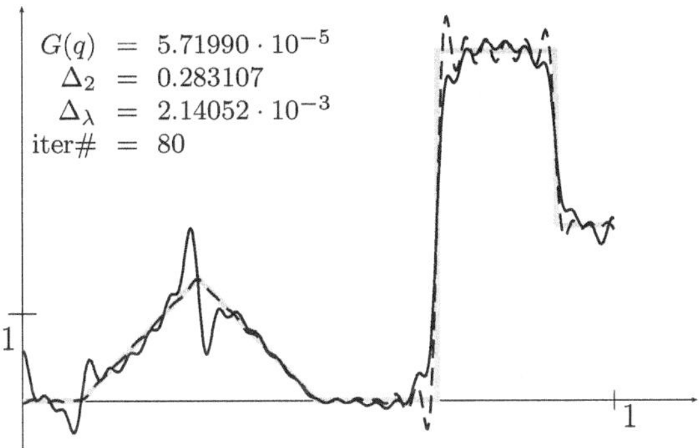

The solid graph is the approximation after 80 iterations and the dashed one after 620 steps. The final result is as good as before, we just need roughly 100 times as many iterations. In the convergence graphs below, we still see the exponential convergence of $G(q)$ to zero.

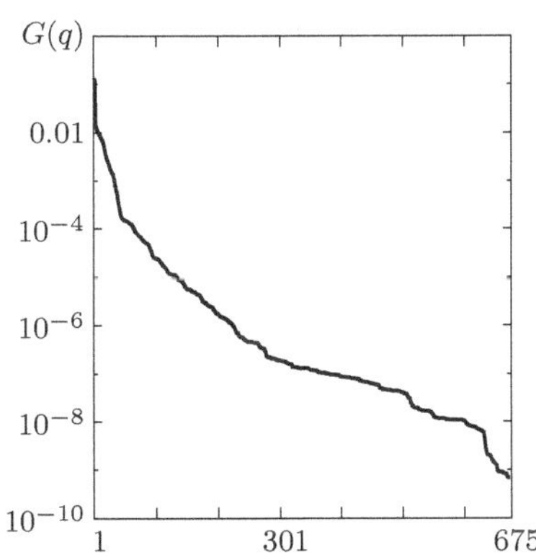

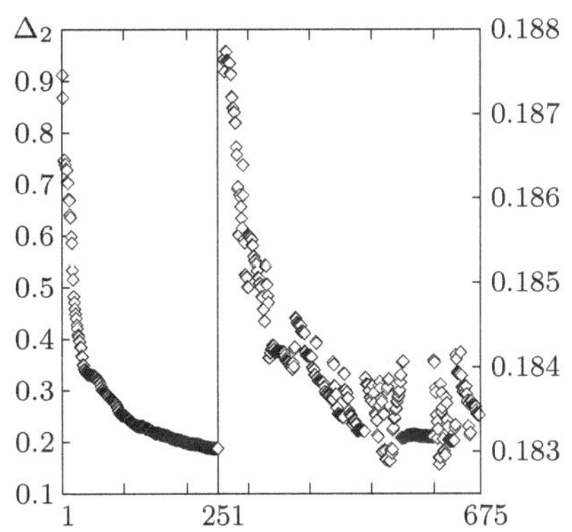

The reason for the slow convergence can be found in the asymptotics of the eigenfunctions. Remember that the gradient of an eigenvalue was given essentially by the square of the corresponding eigenfunction. In the case $\alpha = \beta = 0$, $\gamma = \pi/2$ we have

$$
\begin{aligned}
g_{1,n}^2(x) &= 1 - \cos\left((2n+2)\pi x\right) + O(n^{-1}) \\
g_{2,n}^2(x) &= 1 + \cos\left((2n+1)\pi x\right) + O(n^{-1})
\end{aligned}
$$

and therefore the squares of the n-th eigenfunctions are asymptotically orthogonal. This means that every eigenvalue has its own dimension and they can be independently adjusted by gradient descent.

In the case $\sin\alpha\sin\beta\sin\gamma \neq 0$ we have

$$\begin{aligned} g_{1,n}^2(x) &= 1+\cos\left((2n)\pi x\right)+O(n^{-1}) \\ g_{2,n}^2(x) &= 1+\cos\left((2n)\pi x\right)+O(n^{-1}) \end{aligned}$$

which says that the squares of the n-th eigenfunctions are asymptotically equal. So the two spectra have to fight each other for each n.

For practical purposes it is also important, that we get good results with noisy data. In the first example, we add uniform noise to each eigenvalue, $\tilde{\lambda} = \lambda + U(-0.01, 0.01)$.

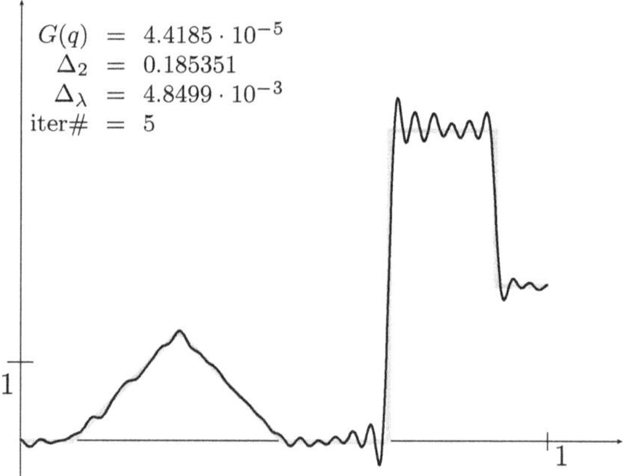

The results are still almost as good as without noise. Increasing the noise by a factor of 10 still yields essentially the right solution.

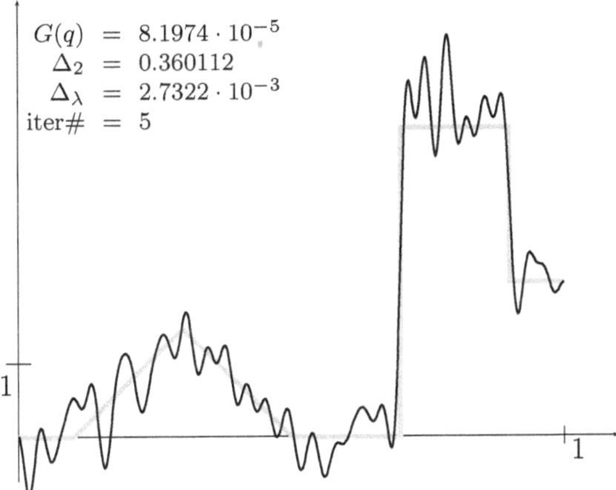

Finally, let us look at a smooth example.

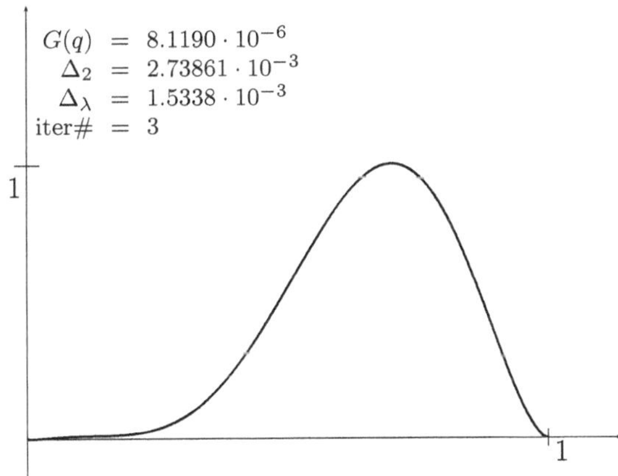

3 Recovering boundary conditions

The spectra also contain information about the boundary conditions. We now show that our method also can recover them.

We exclude the case of Dirichlet ($\alpha = 0$) boundary conditions, since they are a singular case in the continuum of all possible values. Therefore, we choose a more convenient formulation of the boundary conditions,

$$h_0 u(0) + u'(0) = 0, \quad h_1 u(1) + u'(1) = 0\,, \quad \boldsymbol{q} := (h_0, h_1, h_2, q)\,.$$

The corresponding eigenvalues satisfy the asymptotic formula

$$\lambda_n = \pi^2 n^2 + 2(h_1 - h_0) + \int_0^1 q(s)\,\mathrm{d}s + a_n\,,$$

where $(a_n) \in l^2$. It can be shown, that two sequences of eigenvalues corresponding to boundary conditions $(h_0 h_1)$ and $(h_0 h_2)$ with $h_1 \neq h_2$ uniquely determine the boundary conditions (Levitan, 1987).

We use the following abbreviations to denote the potential including the boundary conditions

$$\boldsymbol{q_1} := (h_0, h_1, q)\,, \quad \boldsymbol{q_2} := (h_0, h_2, q)\,, \quad \boldsymbol{q} := (h_0, h_1, h_2, q)\,.$$

Basically we just have to add the boundary information to our functional and compute the full gradient. This leads to (Isaacson and Trubowitz, 1983)

Functional $\quad G(\boldsymbol{q}) := \sum_{(i,n)\in I} (\lambda_{\boldsymbol{q},i,n} - \lambda_{\boldsymbol{Q},i,n})^2$

Gradient of eigenvalues $\quad \nabla\lambda_{\boldsymbol{q},i,n} = \begin{pmatrix} \frac{\partial\lambda_{\boldsymbol{q},i,n}}{\partial h_0} \\ \frac{\partial\lambda_{\boldsymbol{q},i,n}}{\partial h_1} \\ \frac{\partial\lambda_{\boldsymbol{q},i,n}}{\partial h_2} \\ \frac{\partial\lambda_{\boldsymbol{q},i,n}}{\partial q} \end{pmatrix} = \begin{pmatrix} -g^2_{\boldsymbol{q},i,n}(0) \\ g^2_{\boldsymbol{q},i,n}(1)\delta_{i,1} \\ g^2_{\boldsymbol{q},i,n}(1)\delta_{i,2} \\ g^2_{\boldsymbol{q},i,n}(x) \end{pmatrix}$

Gradient $\quad \nabla G(\boldsymbol{q}) = 2 \sum_{(i,n)\in I} (\lambda_{\boldsymbol{q},i,n} - \lambda_{\boldsymbol{Q},i,n}) \nabla\lambda_{\boldsymbol{q},i,n}$

As before, we can also in this case prove the absence of local minima and exponential convergence of the functional.

3.1 Examples

These examples were first published by the American Mathematical Society in (Röhrl, 2006). The first example tries to recover the potential Q with true boundary conditions $H_0 = 3, H_1 = 3, H_2 = 0$ from $q_0 = 0$ and $h_0 = 2, h_1 = 4, h_2 = -1$. The solid graph is the approximation after 150 iterations and the dashed one after 900.

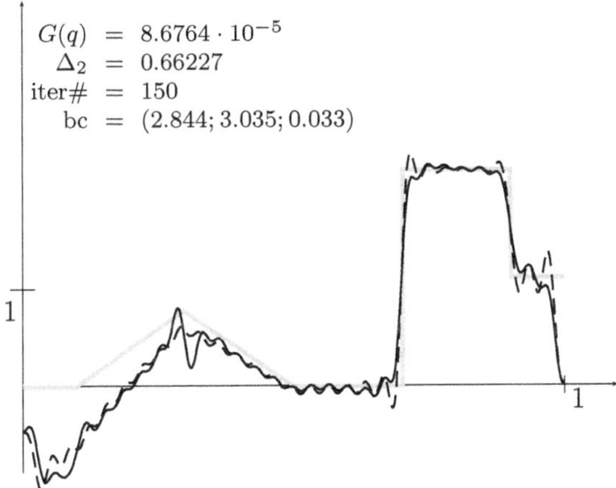

Although the approximation does get better with more iterations, we see in the graphs below, that the boundary conditions do not change after 150 iterations.

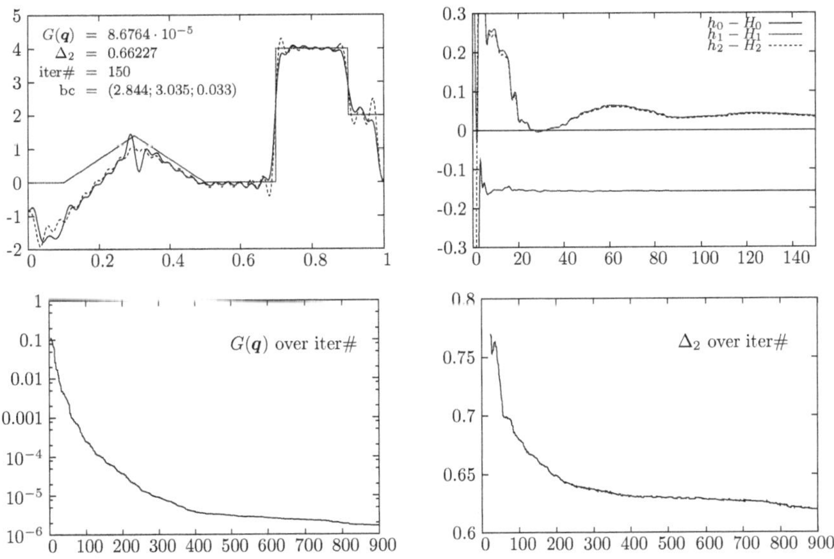

This observations leads to the idea, that we could set q to zero again, once the boundary conditions stabilize. Below we set $q^{(30)} = q^{(60)} = q^{(110)} = 0$ and do get a significantly better approximation in 350 iterations.

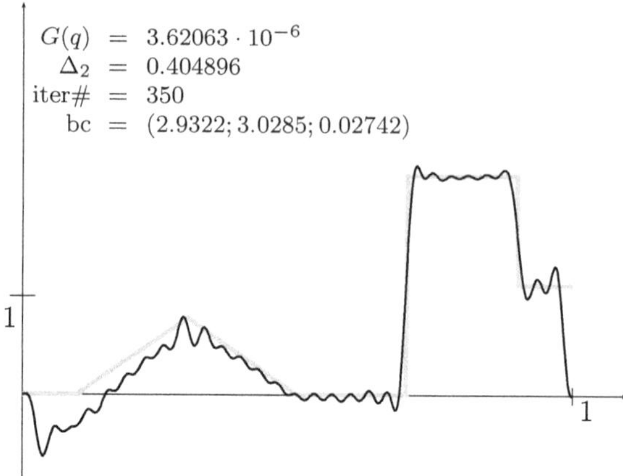

If we start from $h_0 = 0, h_1 = 0, h_2 = 0$ and set $q = 0$ a couple of times, the algorithm still converges.

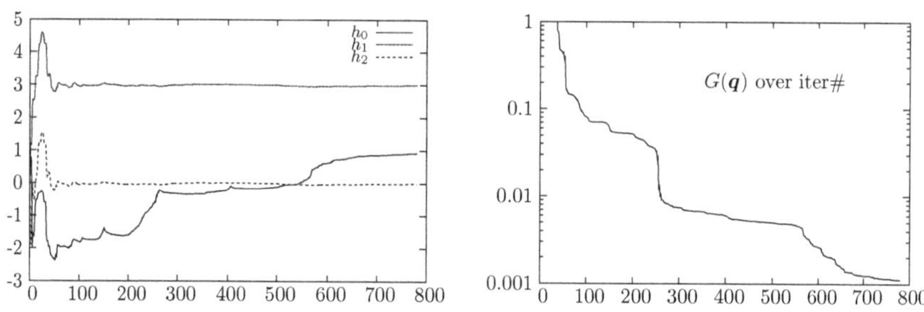

Noisy input data leads to instability. The solid graph after 237 iterations is actually a better approximation than the dashed after 389. We again set $q = 0$ at $i = 89$ to find better solutions.

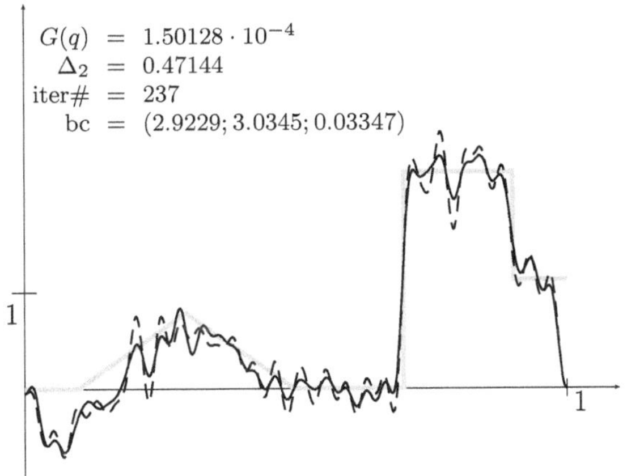

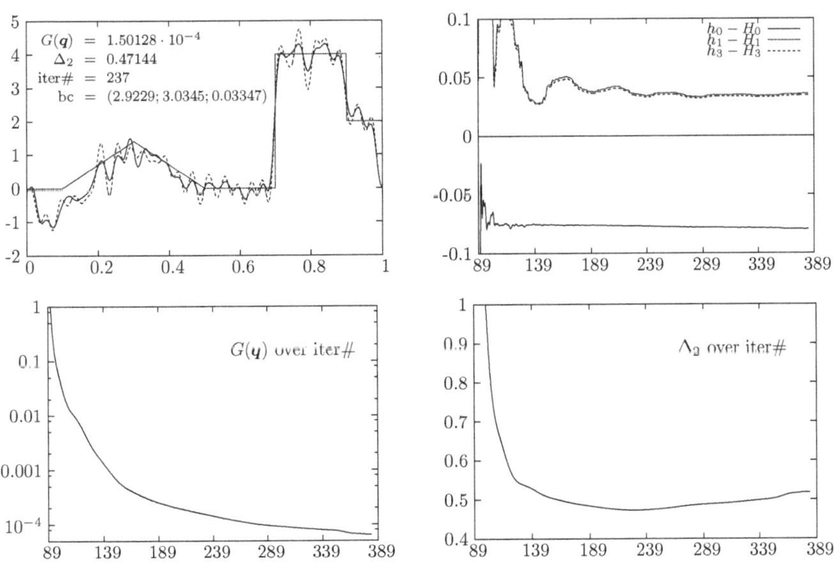

Without setting $q = 0$, the approximation would deteriorate much faster.

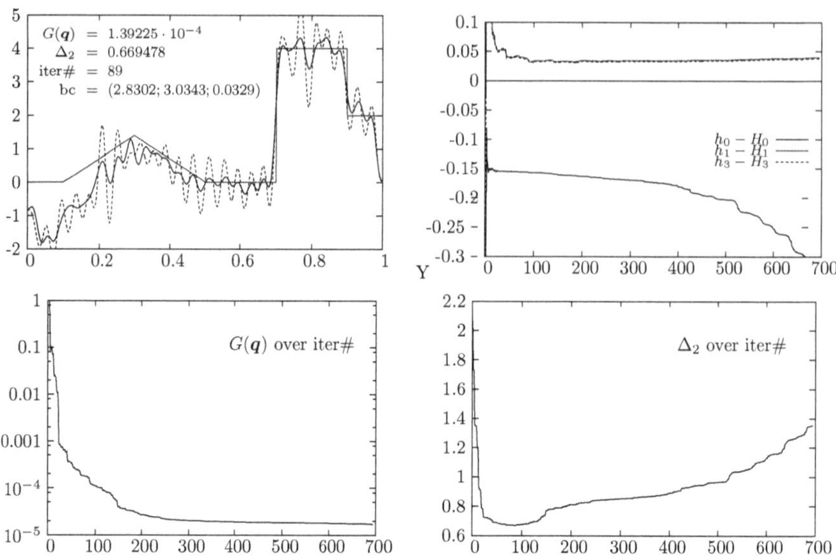

Finally, again the continuous example.

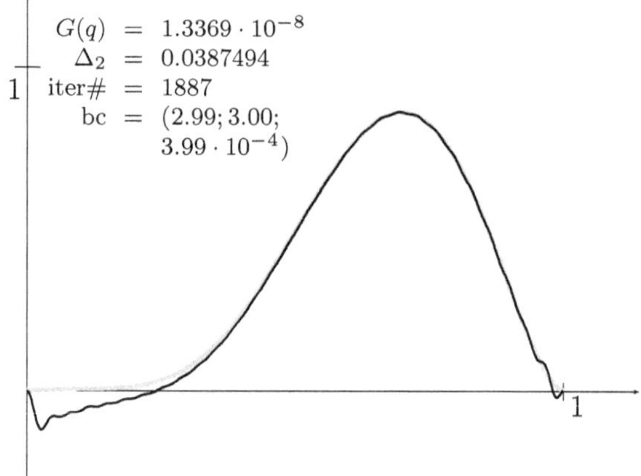

4 Theory

In this section we want to give a quick idea of why this approach works so smoothly.

4.1 Manifolds

A manifold is a space which is locally homeomorphic to an open subset of a Banach space. This means in finite dimensions, that it is locally homeomorphic to an open ball in $\mathbb{R}^n$. For this short introduction we will only consider finite dimensional manifolds.

Let $f : \mathbb{R}^n \to \mathbb{R}^m$ be a differentiable function with derivative $\mathrm{d}f_x : \mathbb{R}^n \to \mathbb{R}^m$ for all $x \in \mathbb{R}^n$. The derivative is a linear map and if it is onto, we can decompose

$$\mathbb{R}^n = T_x \oplus N_x \,,$$

where $\mathrm{d}f_x|_{T_x} = 0$ and $\mathrm{d}f_x|_{N_x} : N \to \mathbb{R}^m$ is a linear isomorphism. We call T_x the tangential and N_x the normal space at point $x \in \mathbb{R}^n$.

Theorem 4.1. *For a constant $c \in \mathbb{R}^m$ the set $M_c = \{x \in \mathbb{R}^n | f(x) = c\}$ is a submanifold of $\mathbb{R}^n$ if $\mathrm{d}f_x$ is onto for all $x \in M_c$.*

Proof. Straight forward application of the implicit function theorem. See also Pöschel and Trubowitz (1987) □

Example 4.2. Let $f : \mathbb{R}^3 \to \mathbb{R}^2, (x_1, x_2, x_3) \mapsto \begin{pmatrix} x_1^2 + x_2^2 + x_3^2 \\ x_3 \end{pmatrix}$, $c = \begin{pmatrix} 1 \\ 0 \end{pmatrix}$. Then the derivative of f,

$$\mathrm{d}f_x : \mathbb{R}^3 \to \mathbb{R}^2, y \mapsto A_x y \,, \quad A_x = \begin{pmatrix} 2x_1 & 2x_2 & 2x_3 \\ 0 & 0 & 1 \end{pmatrix},$$

is onto for all $x \in M_c$, since $x_3 = 0$ and $x_1^2 + x_2^2 = 1$. The decomposition in tangential and normal space is

$$\begin{aligned} T_x &= \lambda \begin{pmatrix} -x_2 \\ x_1 \\ 0 \end{pmatrix}, \\ N_x &= \lambda_1 \begin{pmatrix} 0 \\ 0 \\ 1 \end{pmatrix} + \lambda_2 \begin{pmatrix} x_1 \\ x_2 \\ 0 \end{pmatrix}. \end{aligned}$$

Lemma 4.3. *If $q : [0, T] \to \mathbb{R}^n$, $q(0) \in M_c$ is a curve such that*

$$\frac{\mathrm{d}q(t)}{\mathrm{d}t} \in T_{q(t)} \,, \quad \forall t \in [0, T]$$

then $q(t) \in M_c$ for all $t \in [0, T]$. In other words, if a curve starts on M_c and its derivative lies in the tangent space for all times, then the curve will stay in the manifold M_c.

Proof. This is just a straight forward application of the chain rule:

$$\frac{\mathrm{d}}{\mathrm{d}t}\left(f \circ q(t)\right) = \mathrm{d}f_{q(t)}\frac{\mathrm{d}q(t)}{\mathrm{d}t} = 0\,.$$

Thus f is constant on the curve q. □

4.2 From the Wronskian to a Scalar Product

The Wronskian is an important tool for dealing with second order linear differential equations. Since we want to work in a geometric setting, we write it in form of a scalar product.

If f and h are two solutions of a second order linear differential equation, the Wronskian of f and h is given by

$$W(f,h) = fh' - f'h\,.$$

Integration by parts of which yields

$$\int_0^1 W(f,h)\,\mathrm{d}x = f(1)h(1) - f(0)h(0) - 2\int_0^1 hf'\,\mathrm{d}x\,.$$

This in turn is the scalar product of the vectors

$$\begin{pmatrix} -h(1) \\ h(0) \\ h \end{pmatrix} \quad \begin{pmatrix} -f(1) \\ -f(0) \\ -2f' \end{pmatrix}$$

in $\mathbb{R}^2 \times L^2[0,1]$

4.3 Linear Independence of the Eigenfunctions

Using this scalar product for both boundary conditions simultaneously, we can show the linear independence of the squared eigenfunctions.

Let $s_{q,i,n}, c_{q,i,n}$ be initial value solutions of the Sturm-Liouville equation

$$-u'' + qu = \lambda_{q,i,n}u\,, \quad \begin{array}{rcl} s_{q,i,n}(1) &=& 1\,, \\ s'_{q,i,n}(1) &=& -h_1\,, \end{array} \quad \begin{array}{rcl} c_{q,i,n}(1) &=& 1\,, \\ c'_{q,i,n}(1) &=& -h_2\,, \end{array}$$

Using the gradient of the eigenvalues and the vectors $V_{q,i,n}$ and $U_{q,i,n}$

$$\nabla\lambda_{q,i,n} = \begin{pmatrix} -g^2_{q,i,n}(0) \\ g^2_{q,i,n}(1)\delta_{i,1} \\ g^2_{q,i,n}(1)\delta_{i,2} \\ g^2_{q,i,n}(x) \end{pmatrix}, V_{q,i,n} = \begin{pmatrix} -g^2_{q,i,n}(0) \\ -g^2_{q,i,n}(1) \\ -g^2_{q,i,n}(1) \\ 2\,\mathrm{d}_x g^2_{q,i,n}(x) \end{pmatrix},$$

$$U_{q,i,n} = \begin{pmatrix} s_{q,i,n}c_{q,i,n}(0) \\ s_{q,i,n}c_{q,i,n}(1) \\ s_{q,i,n}c_{q,i,n}(1) \\ -2\,\mathrm{d}_x s_{q,i,n}c_{q,i,n} \end{pmatrix},$$

one can prove the following theorem (Isaacson and Trubowitz, 1983)

Theorem 4.4. *For all $i, j \in \{1, 2\}$ and $n, m \in \mathbb{N}$*

$$\langle \nabla\lambda_{q,i,n}, V_{q,j,m}\rangle = 0\,, \langle \nabla\lambda_{q,i,n}, U_{q,j,m}\rangle = \delta_{ij}\delta_{mn}(-1)^i(h_2 - h_1)\,.$$

This immediately implies the linear independence of the gradients of the eigenvalues.

Theorem 4.5. *The set $\{\nabla\lambda_{q,i,n}|(i,n) \in \{1,2\} \times \mathbb{N}\}$ is linearly independent*

Proof. Suppose

$$0 = \sum_{(i,n)} \alpha_{i,n}\nabla\lambda_{q,i,n}\,, \quad \alpha_{(i,n)} \in \mathbb{R}\,.$$

Since the scalar product is continuous we can pull the sum in front and get for all (j, m)

$$\begin{aligned} 0 &= \langle \sum_{(i,n)} \alpha_{i,n}\nabla\lambda_{q,i,n}, U_{q,j,m}\rangle \\ &= \sum_{(i,n)} \alpha_{i,n}\langle \nabla\lambda_{q,i,n}, U_{q,j,m}\rangle = \alpha_{j,m}(-1)^j(h_2 - h_1) \end{aligned}$$

□

This allows us to see that all potentials which produce our finitely many eigenvalues lie on a manifold. To that end fix a finite set of eigenvalues $\{\lambda_{Q,i,n}|(i,n) \in I \subseteq \{1,2\} \times \mathbb{N}\}$ and the map

$$f : \mathbb{R}^3 \times L^2[0,1] \to \mathbb{R}^{|I|}\,, \quad \mathbf{q} = (h_0, h_1, h_2, q) \mapsto (\lambda_{\mathbf{q},i,n})_{(i,n)\in I}$$

whose derivative

$$\mathrm{d}f_{\mathbf{q}} : \mathbb{R}^3 \times L^2[0,1] \to \mathbb{R}^{|I|}\,, \quad \tilde{\mathbf{q}} \mapsto \left(\int_0^1 \nabla\lambda_{q,i,n}\tilde{\mathbf{q}}\,\mathrm{d}x\right)_{(i,n)\in I}$$

is onto everywhere. The set $M = \{\mathbf{q} = (h_0, h_1, h_2, q)|f(\mathbf{q}) = (\lambda_{Q,i,n})_{(i,n)\in I}\}$ therefore is a manifold and $\{\nabla\lambda_{\mathbf{q},i,n}|(i,n) \in I\}$ is a basis of the normal space at $\mathbf{q}$.

4.4 Exponential Convergence

At first we look at general properties of the gradient flow of an arbitrary functional G

$$\frac{\mathrm{d}q_t}{\mathrm{d}t} = -\nabla G(q)$$

Suppose it exists on some set K_C. The change of the value of the functional in t is given by

$$\begin{aligned}\frac{\mathrm{d}G(q_t)}{\mathrm{d}t} &= \frac{\partial G(q)}{\partial q}\frac{\mathrm{d}q_t}{\mathrm{d}t} = \int_0^1 \nabla G(q_t)\frac{\mathrm{d}q_t}{\mathrm{d}t}\,\mathrm{d}x \\ &= -\int_0^1 |\nabla G(q_t)|^2\,\mathrm{d}x = -\|\nabla G(q_t)\|_2^2 .\end{aligned}$$

If we can now find a constant $d > 0$ such that $\|\nabla G(q)\|_2^2 \geq dG(q)$ in K_C, we get

$$\frac{\mathrm{d}G(q_t)}{\mathrm{d}t} \leq -dG(q)\,, \qquad \text{hence } G(q_t) \leq G(q_0)\mathrm{e}^{-dt}\,,$$

if q_t does not leave K_C.

If we can moreover find a constant $\hat{d}$ such that $\|\nabla G(q)\|_2^2 \leq \hat{d}G(q)$ on K_C.

$$\|\nabla G(q_t)\|_2^2 \leq \hat{d}G(q_t) \leq \hat{d}G(q_0)\mathrm{e}^{-dt} \qquad \text{on } K_C\,.$$

A short calculation using the Hölder inequality now bounds $\|q_t - q_0\|_2$.

$$\begin{aligned}\|q_t - q_0\|_2^2 &= \int_0^1\left(\int_0^t \nabla G(q_s)\,\mathrm{d}s\right)^2\mathrm{d}x \\ &= \int_0^1\int_0^t\int_0^t \nabla G(q_{s_1})\nabla G(q_{s_2})\,\mathrm{d}s_1\,\mathrm{d}s_2\,\mathrm{d}x \\ &\leq \int_0^t\int_0^t \|\nabla G(q_{s_1})\|_2\|\nabla G(q_{s_2})\|_2\,\mathrm{d}s_1\,\mathrm{d}s_2 \\ &= \left(\int_0^t \|\nabla G(q_x)\|_2\,\mathrm{d}x\right)^2 \leq \left(\int_0^t \sqrt{\hat{d}G(q_x)}\,\mathrm{d}x\right)^2 \\ &\leq \hat{d}G(q_0)\left(\int_0^t \mathrm{e}^{-dx/2}\,\mathrm{d}x\right)^2 \leq \hat{d}\frac{4}{d^2}G(q_0)\end{aligned}$$

Finally, we have the estimate

$$\|q_t - q_0\|_2^2 \leq \hat{d}\frac{4}{d^2}G(q_0) \tag{3}$$

and therefore the maximal distance travelled by the gradient flow is bounded linearly in $G(q_0)$.

To show that the gradient flow and the two constants actually exist in our case, one can show that

$$0 \neq \frac{\|\nabla G(q)\|_2^2}{G(q)}$$

is a weakly continuous function of q. Using that $K_C = \{q | \|q\|_2 \leq C\}$ is weakly compact we can conclude, that on K_C there exist $d, \hat{d}$ as above.

For a fixed $C > \|q_0\|_2$ the gradient flow can not leave K_C by (3) if $G(q_0)$ is small enough. In summary we get the following theorem.

Theorem 4.6. *If $G(q_0)$ is sufficiently small, there exists a $d > 0$ such that*

$$G(q_t) \leq G(q_0)\mathrm{e}^{-dt}$$

for all $t \geq 0$.

4.5 Other spectral Data

There are also uniqueness results when one sequence of eigenvalues $\lambda_{\mathbf{q},n}$ and either one of

1. $\rho_n = \int_0^1 z_n^2 \, dx$, where z_n is a solution of $-u'' + qu = \lambda_{\mathbf{q},n} u, u(0) = -1, u'(0) = h_0$ (Gel'fand and Levitan, 1955).
2. $l_{\mathbf{q},n} = \ln\left(\frac{(-1)^n g_{\mathbf{q},n}(1)}{g_{\mathbf{q},n}(0)}\right)$ (Isaacson and Trubowitz, 1983).
3. C_n (Levinson, 1949) defined as follows: For the two initial value solutions u_λ, v_λ with $u_\lambda(0) = -1, u'_\lambda(0) = h_0$, $v_\lambda(1) = -1, v'_\lambda(1) = h_1$ there is are are constants C_n with $u_\lambda = C_n v_\lambda$ if $\lambda = \lambda_{\mathbf{q},n}$, since then both solutions are multiples of the eigenfunction.

4.6 Isospectral Manifolds

Finally I want to sketch, what (Isaacson and Trubowitz, 1983) showed about the structure of the set of potentials which have one fixed sequence of eigenvalues. First they show, that the sequences $\lambda_{\mathbf{q},n}$ and $l_{\mathbf{q},n}$ uniquely determine the potential including boundary conditions

Theorem 4.7. *The map $\mathbf{q} = (h_0, h_1, q) \mapsto (\lambda_{\mathbf{q},n}, l_{\mathbf{q},n})$ is one-to-one.*

Then they define the isospectral set and show that it is a manifold.

Definition 4.8.

$$M(\mathbf{p}) = \{\mathbf{q} | \lambda_{\mathbf{q},n} = \lambda_{\mathbf{p},n} \forall n \geq 0\}$$

Theorem 4.9. *$M(\mathbf{p})$ is a manifold. A basis of its normal space at point $\mathbf{q}$ is $\{\nabla\lambda_{\mathbf{q},n} | n \in \mathbb{N}\}$ and a basis of the tangent space is $\{V_{\mathbf{q},n}\}$.*

They also explicitly compute the flow

$$\frac{\mathrm{d}\mathbf{q}_n(t)}{\mathrm{d}t} = -V_n(\mathbf{q}_n(t))$$

and show that

$$l_{\mathbf{q}_n(t),m} = \begin{cases} l_{\mathbf{q}_n(0),m} & m \neq n \\ l_{\mathbf{q}_n(0),n} + t & m = n \end{cases}$$

Therefore we have the nice geometrical theorem.

Theorem 4.10. *$(l_{\mathbf{q},n})_{n\in\mathbb{N}}$ is a global coordinate system for $M(\mathbf{p})$ and thus the manifold is connected.*

5 Similar Algorithms

Finally, we list a number of similar algorithms, since no one is optimal for all problems.

- Rundell and Sacks (Rundell and Sacks, 1992) developed an elegant an efficient algorithm based on Gelfand-Levitan-Marchenko kernels. On it's downside it needs the mean $\int_0^1 Q\,\mathrm{d}x$ of the potential and the boundary conditions as additional inputs besides the two spectra.
- Brown, Samko, Knowles, and Marletta (Brown et al., 2003) invented an algorithm using Levinson (Levinson, 1949) spectral data. It does not need the mean as a separate input, but it is unknown if it also can be used to recover the boundary conditions.
- Lowe, Pilant, and Rundell (Lowe et al., 1992) use a finite basis ansatz, and solve the inverse problem by Newton's method without requiring the mean as input.

Bibliography

G. Borg. Eine Umkehrung der Sturm-Liouvilleschen Eigenwertaufgabe. Bestimmung der Differentialgleichung durch die Eigenwerte. *Acta Math.*, 78:1–96, 1946.

B. M. Brown, V. S. Samko, I. W. Knowles, and M. Marletta. Inverse spectral problem for the Sturm-Liouville equation. *Inverse Problems*, 19 (1):235–252, 2003. ISSN 0266-5611.

Björn E. J. Dahlberg and Eugene Trubowitz. The inverse Sturm-Liouville problem. III. *Comm. Pure Appl. Math.*, 37(2):255–267, 1984. ISSN 0010-3640.

I. M. Gel′fand and B. M. Levitan. On the determination of a differential equation from its spectral function. *Amer. Math. Soc. Transl. (2)*, 1: 253–304, 1955.

E. L. Isaacson and E. Trubowitz. The inverse Sturm-Liouville problem. I. *Comm. Pure Appl. Math.*, 36(6):767–783, 1983. ISSN 0010-3640.

N. Levinson. The inverse Sturm-Liouville problem. *Mat. Tidsskr. B.*, 1949: 25–30, 1949.

B. M. Levitan. *Inverse Sturm-Liouville problems.* VSP, Zeist, 1987. ISBN 90-6764-055-7. Translated from the Russian by O. Efimov.

B. D. Lowe, M. Pilant, and W. Rundell. The recovery of potentials from finite spectral data. *SIAM J. Math. Anal.*, 23(2):482–504, 1992. ISSN 0036-1410.

W. H. Press, S. A. Teukolsky, W. T. Vetterling, and B. P. Flannery. *Numerical recipes in C.* Cambridge University Press, Cambridge, second edition, 1992. ISBN 0-521-43108-5. URL `http://www.nrbook.com/a/bookcpdf.php`.

Jürgen Pöschel and Eugene Trubowitz. *Inverse spectral theory*, volume 130 of *Pure and Applied Mathematics*. Academic Press Inc., Boston, MA, 1987. ISBN 0-12-563040-9.

W. Rundell and P. E. Sacks. Reconstruction techniques for classical inverse Sturm-Liouville problems. *Math. Comp.*, 58(197):161–183, 1992. ISSN 0025-5718.

N. Röhrl. A least squares functional for solving inverse Sturm-Liouville problems. *Inverse Problems*, 21:2009–2017, 2005. ISSN 0266-5611.

N. Röhrl. Recovering boundary conditions in inverse Sturm-Liouville problems. In *Recent advances in differential equations and mathematical physics*, volume 412 of *Contemp. Math.*, pages 263–270. Amer. Math. Soc., Providence, RI, 2006.

E. C. Titchmarsh. *Eigenfunction expansions associated with second-order differential equations. Part I.* Second Edition. Clarendon Press, Oxford, 1962.

Boundary Control Method in Dynamical Inverse Problems - An Introductory Course

Mikhail I. Belishev *

Saint-Petersburg Department of the Steklov Mathematical Institute
Fontanka 27, Saint-Petersburg 191023, Russian Federation

1 Introduction

1.1 About the paper

The boundary control method (*BC-method*) is an approach to inverse problems based on their relations to control theory. Originally, it was proposed for solving multidimensional problems (see [2]) and therefore such an approach cannot be simple: the BC-method uses asymptotic methods in PDEs, functional analysis, control and system theory, etc. The aim of this lecture course is to provide possibly elementary and transparent introduction to the BC-method. Namely, we present its 1-dimensional version on example of two dynamical inverse problems.

The first problem is to recover the coefficient $q = q(x)$ in the string equation $\rho u_{tt} - u_{xx} + qu = 0$ with $\rho =$ const > 0 on the semi-axis $x > 0$ via the time-domain measurements at the endpoint $x = 0$ of the string. The problem is classical and well investigated, its versions were solved in 60-70's in the pioneer papers by Blagovestchenskii, Sondhi and Gopinath. So, this part of the course is of educational character; we solve a classical problem but by another technique: the main tools of the BC-method (amplitude formula, wave visualization, etc) are introduced and applied for determination of q.

The second problem is more difficult and richer in content: we deal with a vector dynamical system governed by the beam equation $\rho u_{tt} - u_{xx} + Au_x + Bu = 0$ in the semi-axis $x > 0$ with a constant diagonal 2×2-matrix $\rho =$ diag $\{\rho_1, \rho_2\}$, $0 < \rho_1 < \rho_2$ and variable 2×2-matrices $A(x)$, $B(x)$. Such an equation describes the wave processes in a system, where two different wave modes occur and propagate with different velocities. The modes

*The work is supported by the RFBR grants No. 08-01-00511 and NSh-4210.2010.1.

interact; interaction provides interesting physical effects and in the mean time complicates the picture of waves. An inverse problem is to recover the coefficients $A, B\mid_{x>0}$ via the time-domain measurements at the endpoint $x = 0$ of the beam. We solve this problem by the BC-method, the parallels to the scalar case (string equation) being clarified and made use of. In contrast to the first part of the course, where we "solve a solved problem", the results concerning to two-velocity dynamical systems are close to the state of the art in this area of inverse problems.

1.2 Comment, notation, convention

- About the style. We try to make the course available for the reader interested in applications. By this, we do not formulate lemmas and theorems but mainly derive formulas. However, every time we outline (or give a hint how to get) the rigorous proofs and provide the reader with proper references.
- All functions in the paper are *real.* The following classes of functions are in use:

  the space $C[a, b]$ of continuous functions and the space $C^k[a, b]$ of k times continuously differentiable functions;

  a Hilbert space $L_2(a, b)$ of square summable functions provided with the standard inner product

$$(y, v)_{L_2(a,b)} := \int_b^a y(s)\, v(s)\, ds$$

  and the norm $\|y\|_{L_2(a,b)} := (y, y)^{\frac{1}{2}}_{L_2(a,b)}$;

  the Sobolev class $H^1[a, b]$ of differentiable functions y with derivatives $y' \in L_2(a, b)$.
- Sometimes, to distinguish the time intervals $\alpha < t < \beta$ from the space ones $a < x < b$ we denote the space intervals by $\Omega^{(a,b)} := (a, b)$ and put $\Omega^a := (0, a)$.
- **Convention** All functions depending on time $t \geq 0$ are assumed to be extended to $t < 0$ by zero.

1.3 Acknowledgements

The author is grateful to I.V.Kubyshkin for kind help in computer graphics and A.L.Pestov for useful discussions.

PART I: 1-CHANNEL DYNAMICAL SYSTEM (STRING)

2 Forward problem

2.1 Statement

We deal with an initial boundary value problem of the form

$$\rho u_{tt} - u_{xx} + qu = 0, \qquad x > 0, \ 0 < t < T \tag{1}$$

$$u|_{t=0} = u_t|_{t=0} = 0, \qquad x \geq 0 \tag{2}$$

$$u|_{x=0} = f, \qquad 0 \leq t \leq T, \tag{3}$$

where $q = q(x)$ is a continuous function (*potential*) on $x \geq 0$; $\rho > 0$ is a constant (*density*); the value $t = T < \infty$ is referred to as a final moment; $f = f(t)$ is a *boundary control*; $u = u^f(x,t)$ is a solution. The solution u^f is interpreted as a *wave* initiated by the control f, the wave propagating along the semi-axis (*string*) $x \geq 0$. We call (1)–(3) the problem 1.

In parallel we consider the *problem* $\breve{1}$

$$\rho u_{tt} - u_{xx} = 0, \qquad x > 0, \ 0 < t < T \tag{4}$$

$$u|_{t=0} = u_t|_{t=0} = 0, \qquad x \geq 0 \tag{5}$$

$$u|_{x=0} = f, \qquad 0 \leq t \leq T \tag{6}$$

and refer to (1)–(3) and (4)–(6) as perturbed and unperturbed problems respectively. The solution of the unperturbed problem is denoted by $\breve{u}^f(x,t)$; it can be found explicitly [1]:

$$\breve{u}^f(x,t) = f\left(t - \frac{x}{c}\right), \tag{7}$$

where $c := \frac{1}{\sqrt{\rho}}$ is a *wave velocity*.

We shall illustrate the considerations with the figures, which show controls and waves (at the final moment) on Fig. 1.

2.2 Integral equation and generalized solutions

The main tool for investigating the problem 1 is an integral equation, which is equivalent to (1)–(3). Seeking for the solution in the form $u^f =$

[1] Regarding the values of f for $t - \frac{x}{c} < 0$, recall the Convention accepted: $f|_{t<0} = 0$!

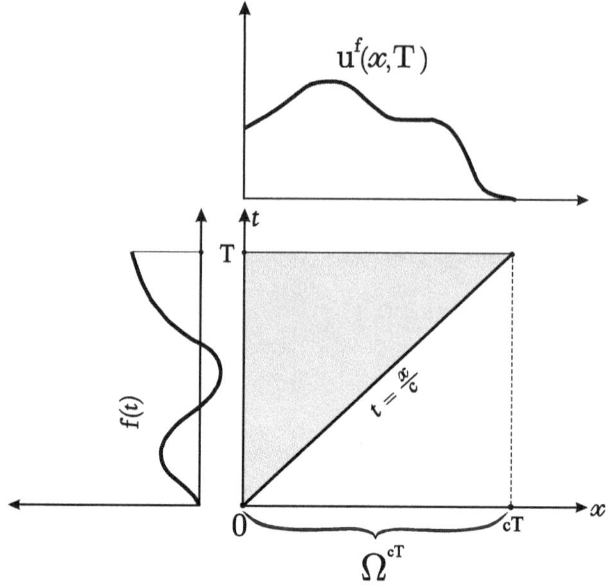

Figure 1. Controls and waves

$\breve{u}^f + w$ with a new unknown $w = w^f(x,t)$, we easily get

$$\rho w_{tt} - w_{xx} = -qw - q\breve{u}^f, \qquad x > 0,\ 0 < t < T \tag{8}$$

$$w|_{t=0} = w_t|_{t=0} = 0, \qquad x \geq 0 \tag{9}$$

$$w|_{x=0} = 0, \qquad 0 \leq t \leq T\,. \tag{10}$$

Applying the D'Alembert formula (see, e.g., [30]), we arrive at the equation

$$w + Mw = -M\breve{u}^f \qquad x > 0,\ 0 \leq t \leq T\,, \tag{11}$$

where an operator M acts by the rule

$$(Mv)\,(x,t) := \frac{1}{2c} \iint\limits_{K_c(x,t)} q(\xi)\, w(\xi,\eta)\, d\xi\, d\eta \tag{12}$$

and $K_c(x,t)$ is the trapezium bounded by the characteristic lines $t \pm \frac{x}{c} =$ const. The equation (11) is a second-kind Volterra type equation; it can be studied by the standard iteration method (see, e.g., [30]). As can be shown, if the control $f \in C^2[0,T]$ satisfies $f(0) = f'(0) = f''(0) = 0$ then

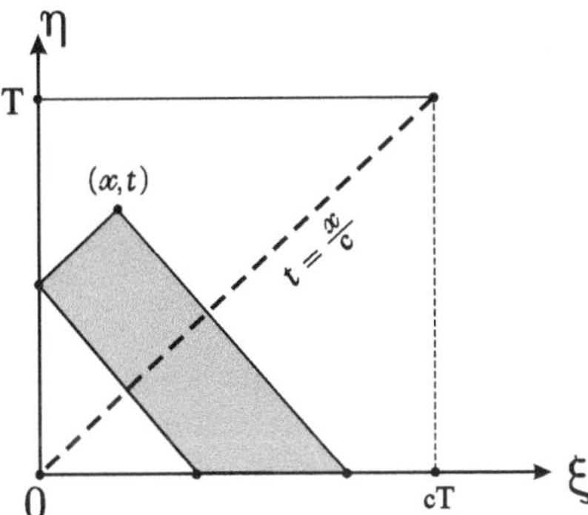

Figure 2. The domain $K_c(x,t)$

its solution w^f is twice continuously differentiable. As a result, the function $u^f = \breve{u}^f + w^f$ turns out to be *classical solution* of the problem 1.

As is easy to see, for any $f \in L_2(0,T)$, the r.h.s. $M\breve{u}^f$ of the equation (11) is a continuous function of x, t vanishing as $t < \frac{x}{c}$. Simple analysis shows that the equation is uniquely solvable in this class of functions. In this case, we define the function $u^f := \breve{u}^f + w^f$ to be a *generalized solution* of the problem 1. So, for any control $f \in L_2(0,T)$, such a solution u^f does exist and is unique.

2.3 Fundamental solution

Denote

$$\delta_\varepsilon(t) := \begin{cases} \frac{1}{\varepsilon}, & 0 \le t < \varepsilon \\ 0, & t \ge \varepsilon \end{cases}$$

and recall that $\delta_\varepsilon \to \delta(t)$ (the Dirac delta-function) as $\varepsilon \to 0$ in the sense of distributions.

Putting $f = \delta_\varepsilon(t)$ in (11) we get the solution w^{δ_ε} belonging to the class $C([0,T]; L_2(a,b))$. In the mean time, letting $\varepsilon \to 0$ we easily see that the r.h.s. of (11) $M\breve{u}^{\delta_\varepsilon}$ tends to $M\breve{u}^\delta$, the latter being a continuously differentiable function of x, t for $0 \le \frac{x}{c} \le t \le T$ and vanishing as $\frac{x}{c} > t$. Thus, even though the limit passage leads to a singular control $f = \delta(t)$, the solution $w^\delta := \lim_{\varepsilon \to 0} w^{\delta_\varepsilon}$ does not leave the class $C([0,T]; L_2(a,b))$. Simple analysis provides the following of its properties:

1. w^δ is continuously differentiable in the domain $\{(x,t) \mid 0 \le x \le cT,\ \frac{x}{c} \le t \le T\}$ and $w^\delta \,|_{t<\frac{x}{c}} = 0$

2. $w^\delta(0,t) = 0$ for all $t \geq 0$
3. the relation

$$w^\delta\left(x-0,\frac{x}{c}\right) = -\frac{c}{2}\int_0^x q(s)\,ds \qquad x \geq 0 \tag{13}$$

holds [2] and shows that w^δ may have a jump at the characteristic line $t = \frac{x}{c}$ (whereas below the line $w^\delta = 0$ holds). We derive (13) later in Appendix I.

In the unperturbed case of $q = 0$, we easily have $\breve{w}^\delta = 0$, whereas

$$\breve{u}^\delta(x,t) = \delta\left(t-\frac{x}{c}\right) \tag{14}$$

satisfies (4) in the sense of distributions and is called a *fundamental solution* of the problem $\breve{1}$. Analogously, the distribution $u^\delta := \breve{u}^\delta + w^\delta$ of the form

$$u^\delta(x,t) = \delta\left(t-\frac{x}{c}\right) + w^\delta(x,t) \tag{15}$$

is said to be a *fundamental solution* of the (perturbed) problem 1. It describes the wave produced by the impulse control $f = \delta(t)$ (see Fig 3).

Such a wave consists of the singular leading part $\delta\left(t-\frac{x}{c}\right)$ propagating along the string with velocity c and the regular *tail* $w^\delta(x,t)$, which may have a jump at its forward front. The presence of the tail is explained by interaction between the singular part and the potential. Also, note that the singular part in the perturbed and unperturbed cases is one and the same [3].

2.4 Properties of waves

Return to the problem 1 and represent the control in the form of a convolution w.r.t. time: $f(t) = (\delta * f)(t) = \int_{-\infty}^{\infty}\delta(t-s)\,f(s)\,ds$. Since the potential in (1) does not depend on time, by the well-known linear superposition principle we have

$$u^f = u^{\delta * f} = u^\delta * f = \left(\breve{u}^\delta + w^\delta\right) * f = \breve{u}^\delta * f + w^\delta * f$$

[2] Here $w^\delta\left(x-0,\frac{x}{c}\right)$ is understood as $\lim_{s\uparrow x} w^\delta\left(s,\frac{x}{c}\right)$.

[3] Physicists comment on this fact as a principle: Sharp signals do not feel a potential.

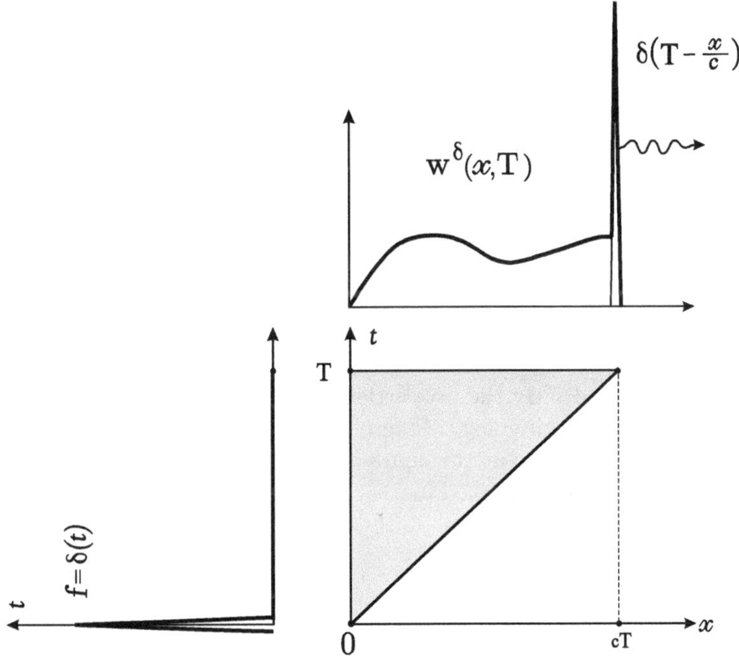

Figure 3. Fundamental solution

that implies the representation

$$u^f(x,t) = f\left(t - \frac{x}{c}\right) + \int_0^t w^\delta(x,t-s)\, f(s)\, ds = \langle \text{see sec 2.3, property 1} \rangle$$

$$= f\left(t - \frac{x}{c}\right) + \int_0^{t-\frac{x}{c}} w^\delta(x,t-s)\, f(s)\, ds\,, \qquad x \geq 0,\ t \geq 0\,, \tag{16}$$

which is often called the Duhamel's formula. The listed below properties of the waves easily follow from (16):

- For any $f \in L_2(0,T)$, the relation

$$u^f\,|_{t<\frac{x}{c}} = 0, \qquad x \geq 0,\ t \geq 0 \tag{17}$$

holds, which is interpreted as the finiteness of the wave propagation speed.

- For $f \in L_2(0,T)$ and $\tau \in [0,T]$, define a delayed control $f_\tau \in L_2(0,T)$ by

$$f_\tau(t) := f(t-\tau)\,, \qquad 0 \leq t \leq T\,,$$

where τ is the value of delay (recall the Convention!). Since the potential does not depend on t, the delay of the control implies the same delay of the wave: the relation

$$u^{f_\tau}(x,t) := u^f(x,t-\tau) \tag{18}$$

is valid.

- Let a control f be a piece-wise continuous function, which has a jump at $t=\xi$ (see Fig 4). By the properties of w^δ, the integral term in (16) depends on x,t continuously. Therefore, $u^f(x,t)$ and $f\left(t-\frac{x}{c}\right)$ have one and the same jump at the characteristic line $t=\frac{x}{c}+\xi$, the values of the jumps being connected as

$$u^f(x,t)\Bigg|_{x=c(t-\xi)-0}^{x=c(t-\xi)+0} = -f(s)\Bigg|_{s=\xi-0}^{s=\xi+0}\,, \qquad t>0. \tag{19}$$

In particular, if f vanishes for $0<t<T-\xi$ and has a jump at

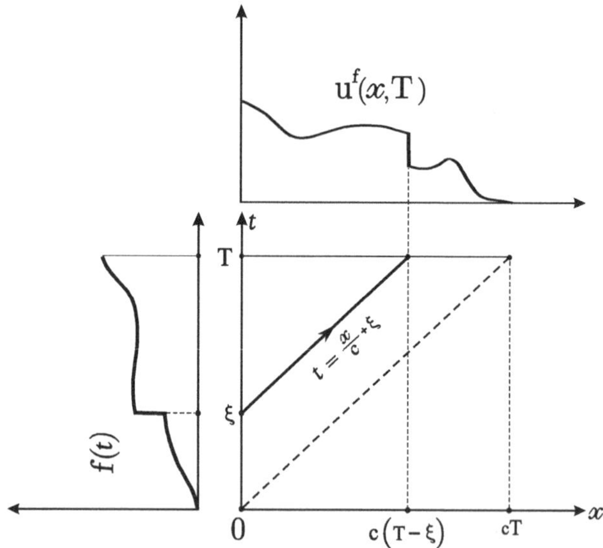

Figure 4. Propagation of jumps

$t = T - \xi$ then $u^f(\,\cdot\,,T)$ vanishes for $x > c\xi$, the amplitudes of the jumps being related through (19) as follows:

$$u^f(c\xi - 0, T) \;=\; f(T - \xi + 0) \qquad 0 < \xi \le T\,. \tag{20}$$

The equalities (19) and (20) is the simplest version of the relations, which are known in PDE-theory as *geometrical optics formulas.* Such formulas describe a propagation of the solution singularities in the wave processes governed by hyperbolic equations.

2.5 Extended problem 1 and locality

Return to the integral equation (11) for the regular part w^δ of the fundamental solution. Representing in the form of the Neumann series $w^\delta = \sum_{k=0}^{\infty}(-1)^k M^k \breve{u}^\delta$ and looking at Fig 2, we easily see that the values $w^\delta(x,t)$ for $0 < x < \frac{cT}{2}$, $t < T - \frac{x}{c}$ (under the line $t = T - \frac{x}{c}$ on Fig 5a) are determined by the values of the potential $q\mid_{0<x<\frac{cT}{2}}$ only (do not depend on the behavior of q in $x \ge \frac{cT}{2}$).

Such a dependence is an inherent feature of the wave processes with the finite speed of the wave propagation; it is known as a *locality principle* (see, e.g., [17]). It motivates to extend the problem 1 as follows:

$$\rho u_{tt} - u_{xx} + qu = 0, \qquad (x,t) \in \Delta^{2T} \tag{21}$$

$$u\mid_{t<\frac{x}{c}} = 0 \tag{22}$$

$$u|_{x=0} = f, \qquad 0 \le t \le 2T\,, \tag{23}$$

where $\Delta^{2T} := \{(x,t)\,|\, 0 < x < cT,\; 0 < t < 2T - \frac{x}{c}\}$ (see the shaded domain on Fig 5b). Seeking for the solution in the form $u^f = f\left(t - \frac{x}{c}\right) + w^f$, one can reduce the problem to the integral equation

$$w + Mw \;=\; -M\breve{u}^f \qquad \text{in } \Delta^{2T}, \tag{24}$$

which is quite analogous to (11) and, for any $f \in L_2(0, 2T)$, is uniquely solvable in a relevant class of functions in Δ^{2T}. So, (21)–(23) is a well-posed problem, its solution being determined by $q\mid_{0<x<cT}$.

Taking in (23) $f = \delta(t)$, one can construct the fundamental solution

$$u^\delta(x,t) = \delta\left(t - \frac{x}{c}\right) + w^\delta(x,t)\,, \qquad (x,t) \in \Delta^{2T}, \tag{25}$$

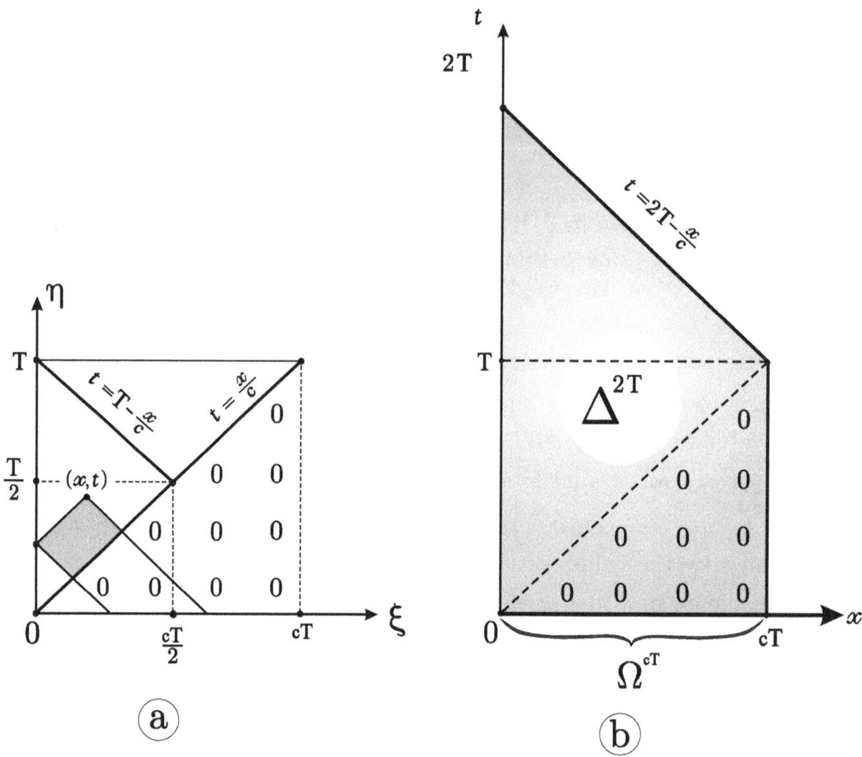

Figure 5. Locality and extended problem

and represent

$$u^f(x,t) = u^\delta(x,t) * f(t) =$$

$$f\left(t-\frac{x}{c}\right) + \int\limits_0^{t-\frac{x}{c}} w^\delta(x,t-s)\, f(s)\, ds\,, \qquad (x,t) \in \Delta^{2T} \qquad (26)$$

that extends the domain of definition of the Duhamel representation (16).

Take a smooth control f provided $f(0)=0$. Differentiating in (26) w.r.t. x and putting $x=0$, we get

$$u^f_x(0,t) = -\sqrt{\rho}\, f'(t) + \int\limits_0^t r(t-s)\, f(s)\, ds\,, \qquad 0<t<2T\,, \qquad (27)$$

where $r(t) := w_x^\delta(0,t)$ is called a *reply function*. It is a smooth function [4]; it will play a central role in the inverse problem.

Remark By the locality principle, the reply function $r\mid_{0<t<2T}$ is determined by the part $q\mid_{0<x<cT}$ of the potential.

3 String as dynamical system

3.1 System α^T

Here problem (1)–(3) is considered in terms of the control theory and endowed with standard attributes of a dynamical system. The system is denoted by α^T.

Outer space The Hilbert space of controls (inputs) $\mathcal{F}^T := L_2(0,T)$ with the inner product

$$(f,g)_{\mathcal{F}^T} = \int_0^T f(t)\,g(t)\,dt$$

is said to be an *outer space* of the system α^T. It contains an increasing family of subspaces

$$\mathcal{F}^{T,\xi} := \{f \in \mathcal{F}^T \mid f|_{t<T-\xi} = 0\}, \qquad 0 \le \xi \le T$$

consisting of the delayed controls ($T-\xi$ is the value of the delay, ξ is an action time). Note that $\mathcal{F}^{T,0} = \{0\}$ and $\mathcal{F}^{T,T} = \mathcal{F}^T$.

Inner space Recall that by Ω^a we denote the interval $(0,a)$ of the semi-axis $x \ge 0$. The space $\mathcal{H}^{cT} := L_{2,\rho}(\Omega^{cT})$ with the inner product

$$(u,v)_{\mathcal{H}^{cT}} = \int_{\Omega^{cT}} u(x)\,v(x)\,\rho\,dx$$

is called an *inner space* of the system α^T. For each $f \in \mathcal{F}^T$, at any moment t, the wave $u^f(\,\cdot\,,t)$ is supported in Ω^{cT} (see (17)) and, hence, can be regarded as a time-dependent element of $\mathcal{H}^{cT}$. In control theory, $u^f(\,\cdot\,,t)$ is referred to as a *state* of the system at the moment t. So, $\mathcal{H}^{cT}$ is a space of states.

The inner space contains an increasing family of subspaces

$$\mathcal{H}^{c\xi} := \{y \in \mathcal{H}^{cT} \mid y|_{x>c\xi} = 0\}, \qquad 0 \le \xi \le T\,.$$

[4] The case $r(+0) \ne 0$ is possible, i.e., the reply function may have a jump at $t=0$ only.

Control operator The input $\mapsto$ state correspondence of the system α^T is realized by a *control operator* $W^T : \mathcal{F}^T \to \mathcal{H}^{cT}$,

$$W^T f := u^f(\,\cdot\,, T)$$

which creates the waves. Putting $t = T$ in (16) and changing the variable in the integral, we get a representation

$$\left(W^T f\right)(x) = f\left(T - \frac{x}{c}\right) + \int_x^{cT} w(x,s)\, f\left(T - \frac{s}{c}\right) ds\,, \qquad x \in \Omega^{cT}\,, \quad (28)$$

where $w(x,s) := \frac{1}{c}\, w^\delta(x, \frac{s}{c})$. As is easy to see, W^T is a bounded operator.

Recall that the delay of controls f_τ is defined in sec 2.4. Introduce the delay operation $D^{T,\xi} : f \mapsto f_{T-\xi}$ as an operator in the outer space $\mathcal{F}^T$ and note that $D^{T,\xi}\mathcal{F}^T = \mathcal{F}^{T,\xi}$. Applying (18) for $\tau = T - \xi,\ t = T$, we get

$$W^T D^{T,\xi} f = u^f(\,\cdot\,,\xi)\,. \qquad (29)$$

Since $D^{T,\xi} f \in \mathcal{F}^{T,\xi}$, the wave $u^f(\,\cdot\,,\xi)$ is supported in $\Omega^{c\xi}$ (see (17) for $t = \xi$), i.e., the inclusion $u^f(\,\cdot\,,\xi) \in \mathcal{H}^{c\xi}$ holds. Hence, (29) implies the embedding

$$W^T \mathcal{F}^{T,\xi} f \subseteq \mathcal{H}^{c\xi}\,, \qquad 0 \le \xi \le T\,. \qquad (30)$$

As we shall see later in sec 3.3, "$\subseteq$" can be replaced by "$=$".

Response operator An input $\mapsto$ output correspondence in the system α^T is realized by a *response operator* $R^T : \mathcal{F}^T \to \mathcal{F}^T$ defined on controls f of the Sobolev class $H^1[0,T]$ provided $f(0) = 0$.

$$\left(R^T f\right)(t) := u^f_x(0,t), \quad 0 < t < T.$$

In mechanics, $u^f_x(0,t)$ is interpreted as a value (at the moment t) of the force generated at the endpoint $x = 0$ of the string by the wave process initiated by the control f.

Differentiating in (16), taking into account the property 2 in sec 2.3, and recalling the definition of the reply function, we arrive at the representation

$$\left(R^T f\right)(t) = -\frac{1}{c}\, f'(t) + \int_0^t w^\delta_x(0, t-s)\, f(s)\, ds =$$

$$-\sqrt{\rho}\, f'(t) + \int_0^t r(t-s)\, f(s)\, ds\,, \qquad 0 < t < T\,. \qquad (31)$$

The presence of differentiation in (31) renders R^T to be an unbounded operator. Its adjoint $\left(R^T\right)^*$ is also unbounded. As is easy to show, it is defined on the controls f of the Sobolev class $H^1[0,T]$ provided $f(T)=0$ and acts by the rule

$$\left(\left(R^T\right)^* f\right)(t) = \sqrt{\rho}\, f'(t) + \int_t^T r(s-t)\, f(s)\, ds\,, \qquad 0<t<T\,. \tag{32}$$

Return to the extended problem 1 and associate with it the *extended response operator* R^{2T} acting in the space $\mathcal{F}^{2T} := L_2(0,2T)$ on controls f of the class $H^1[0,2T]$ provided $f(0)=0$ by the rule

$$\left(R^{2T} f\right)(t) := u^f_x(0,t), \quad 0<t<2T\,,$$

where u^f is the solution of (21)–(23). Quite analogously to (31), it can be represented in the form

$$\left(R^{2T} f\right)(t) = -\sqrt{\rho}\, f'(t) + \int_0^t r(t-s)\, f(s)\, ds\,, \qquad 0<t<2T\,. \tag{33}$$

The extended response operator is one more intrinsic object of the system α^T. By the locality principle, R^{2T} **is determined by the part** $q\,|_{\Omega^{cT}}$ **of the potential** [5].

Connecting Operator Since the control operator W^T acts from a Hilbert space $\mathcal{F}^T$ to a Hilbert space $\mathcal{H}^{cT}$, its adjoint $\left(W^T\right)^*$ acts from $\mathcal{H}^{cT}$ to $\mathcal{F}^T$. By this, an operator

$$C^T := \left(W^T\right)^* W^T$$

is well defined. It acts in the outer space $\mathcal{F}^T$ and is said to be a *connecting operator* of the system α^T. By continuity of W^T, the operator C^T is also continuous. By the definition, one has

$$\begin{aligned}\left(C^T f, g\right)_{\mathcal{F}^T} &= \left(\left(W^T\right)^* W^T f, g\right)_{\mathcal{F}^T} = \left(W^T f, W^T g\right)_{\mathcal{H}^{cT}} \\ &= \left(u^f(\cdot,T), u^g(\cdot,T)\right)_{\mathcal{H}^{cT}}\,,\end{aligned} \tag{34}$$

i.e., C^T connects the metrics of the outer and inner spaces.

[5] See the remark at the end of sec 2.5 .

The connecting operator plays a key role in the BC-method owing to the following remarkable fact: it can be represented via the response operator in explicit and simple form. To formulate the result we need to introduce auxiliary operators:

- The operator $S^T : \mathcal{F}^T \to \mathcal{F}^{2T}$,

$$\left(S^T f\right)(t) := \begin{cases} f(t)\,, & 0<t<T \\ -f(2T-t)\,, & T \le t < 2T \end{cases}$$

extending the controls from $(0,T)$ to $(0,2T)$ by oddness w.r.t. $t=T$. As is easy to check, its adjoint $\left(S^T\right)^* : \mathcal{F}^{2T} \to \mathcal{F}^T$ acts by the rule

$$\left(\left(S^T\right)^* g\right)(t) \;=\; g(t) - g(2T-t)\,, \qquad 0<t<T\,. \tag{35}$$

- The integration operator $J^{2T} : \mathcal{F}^{2T} \to \mathcal{F}^{2T}$,

$$\left(J^{2T} f\right)(t) = \int\limits_0^t f(\eta)\,d\eta, \quad 0<t<2T\,.$$

The integration commutes with the response operator:

$$R^{2T} J^{2T} \;=\; J^{2T} R^{2T}; \tag{36}$$

this easily follows from the representation (33).

The formula, which expresses the connecting operator via the response operator, is

$$C^T \;=\; -\frac{1}{2}\,\left(S^T\right)^* R^{2T} J^{2T} S^T\,; \tag{37}$$

its derivation see in Appendix I. Furthermore, substituting (33) to (37) one can represent C^T through the reply function:

$$\left(C^T f\right)(t) \;=\; \sqrt{\rho}\, f(t) + \int\limits_0^T c^T(t,s)\, f(s)\, ds\,, \qquad 0<t<T\,, \tag{38}$$

with the kernel

$$c^T(t,s) \;:=\; \frac{1}{2} \int\limits_{|t-s|}^{2T-t-s} r(\eta)\, d\eta \qquad 0<s,t<T\,.$$

To derive (38) from (37) is a simple exercise.

The following fact will be used later in solving the inverse problem. Assume that the external observer investigates the system α^T via its input $\mapsto$ output correspondence. Such an observer operates at the endpoint $x = 0$ of the string; he can apply controls f and measure $u^f_x\,|_{x=0}$ but, however, cannot see the waves u^f themselves on the string [6]. As result of such measurements, the observer is provided with the reply function $r\,|_{(0,2T)}$. If so, the observer can determine the operator C^T by (38) and then, for any given controls $f, g \in \mathcal{F}^T$, find the product of the waves $\big(u^f(\,\cdot\,,T), u^g(\,\cdot\,,T)\big)_{\mathcal{H}^{cT}}$ by (34), even though the waves themselves are invisible! As we shall see, such an option enables one to make the waves visible.

In conclusion, we present the main objects of the system α^T on the diagram:

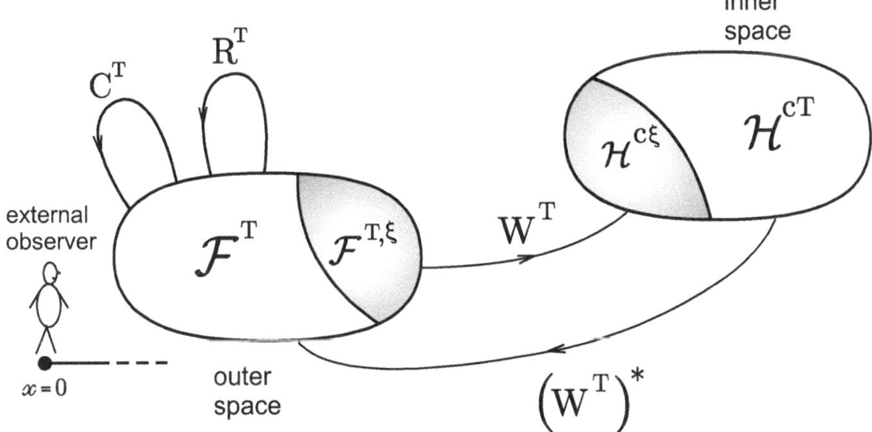

Figure 6. System α^T

3.2 Controllability

The question, which we consider here, can be posed as follows: Can one manage the shape of the wave on a string? More precisely: Is it possible to drive the system α^T from the initial zero state (see (2)) to a given final state $u^f(\,\cdot\,,T) = y$ by means of the proper choice of the boundary control f? This sort of problems is a subject of the boundary control theory, which is a highly developed branch of mathematical physics (see, e.g., [26], [29]). Here the affirmative answer to the posed question is provided.

[6] that is a rule of game in inverse problems!

There is an evident necessary condition for the above-mentioned problem to be solvable. In view of the property (17), the function y has to be supported in the interval Ω^{cT} filled with waves at the final moment. Therefore, the relevant setup is the following: *given function* $y \in \mathcal{H}^{cT}$, *to find control* $f \in \mathcal{F}^T$ *such that*

$$u^f(x,T) = y(x)\,, \qquad x \in \Omega^{cT} \tag{39}$$

holds. It is what is called a *boundary control problem* (BCP). The BCP is obviously equivalent to the equation

$$W^T f = y\,. \tag{40}$$

In the case of the unperturbed system, by (7) the BCP takes the form of the equation

$$f\left(T - \frac{x}{c}\right) = y(x)\,, \qquad x \in \Omega^{cT}\,,$$

which has the evident solution

$$f(t) = y\left(c(T-t)\right)\,, \qquad 0 < t < T\,. \tag{41}$$

In the perturbed case, changing the variables in (28) we get

$$f(t) + \int_0^t k^T(t,s)\, f(s)\, ds = y\left(c(T-t)\right)\,, \qquad 0 < t < T \tag{42}$$

with $k^T(t,s) := w\left(c(T-t), c(T-s)\right)$ that is a second-kind Volterra equation w.r.t. f. As is well known (see, e.g., [30]), such an equation (and hence the BCP) is uniquely solvable in $\mathcal{F}^T$ for any r.h.s. In operator terms, this means that the control operator is boundedly invertible, its inverse $\left(W^T\right)^{-1}$ being defined onto $\mathcal{H}^{cT}$.

This result can be also interpreted as follows. The set of waves

$$\mathcal{U}^T := \{u^f(\,\cdot\,,T) \mid f \in \mathcal{F}^T\} = W^T \mathcal{F}^T$$

is called *reachable* (at the moment $t = T$). The solvability of the BCP is equivalent to the relation

$$\mathcal{U}^T = \mathcal{H}^{cT}\,, \qquad T \geq 0\,, \tag{43}$$

which shows that our system can be steered from zero state for *any* state by proper choice of the boundary control. In control theory, such a property of a dynamical system is referred to as a controllability. Taking into account the finiteness of the wave propagation speed, we specify it as a *local boundary controllability* of the system α^T.

For inverse problems, controllability is an affirmative and helpful property. A very general principle of *system theory* claims that the richer the set of states, which the observer can create in the system by means of the given reserve of controls, the richer information about the system, which the observer can extract from external measurements [22]. As we shall see, the BC-method follows and realizes this principle.

3.3 Wave basis

Here we make use of controllability of the system α^T for storing up an efficient instrument, which will be used for solving the inverse problem.

Fix a positive $\xi \le T$. Since $T > 0$ in (43) is arbitrary, one can put $t = T$ and $\tau = T - \xi$ in (18) and rewrite the controllability relation in the form

$$W^T \mathcal{F}^{T,\,\xi} = \mathcal{H}^{c\xi}, \qquad 0 \le \xi \le T. \tag{44}$$

Let us choose a basis of controls $\{f_j^\xi\}_{j=1}^\infty$ in the subspace $\mathcal{F}^{T,\,\xi}$. Since W^T is a boundedly invertible operator, the relation (44) yields that the corresponding waves $\{u^{f_j^\xi}(\,\cdot\,,T)\}_{j=1}^\infty$ constitute a basis of the subspace $\mathcal{H}^{c\xi}$. For the needs of the inverse problem, it is convenient to make the latter basis orthonormalized. To this end, we recast the basis of controls by the Schmidt orthogonalization process w.r.t. the bilinear form $(C^T f, g)_{\mathcal{F}^T}$:

$$
\begin{aligned}
&\tilde g_1^\xi := f_1^\xi, && g_1^\xi := \frac{\tilde g_1^\xi}{\sqrt{\left(C^T \tilde g_1^\xi,\, \tilde g_1^\xi\right)_{\mathcal{F}^T}}};\\
&\tilde g_2^\xi := f_2^\xi - \left(C^T f_2^\xi,\, g_1^\xi\right)_{\mathcal{F}^T} g_1^\xi, && g_2^\xi := \frac{\tilde g_1^\xi}{\sqrt{\left(C^T \tilde g_2^\xi,\, \tilde g_2^\xi\right)_{\mathcal{F}^T}}};\\
&\cdots\cdots\cdots\cdots\cdots\cdots\cdots\cdots\cdots\cdots\cdots\cdots\\
&\tilde g_k^\xi := f_k^\xi - \sum_{j=1}^{k-1} \left(C^T f_k^\xi,\, g_j^\xi\right)_{\mathcal{F}^T} g_j^\xi, && g_k^\xi := \frac{\tilde g_k^\xi}{\sqrt{\left(C^T \tilde g_k^\xi,\, \tilde g_k^\xi\right)_{\mathcal{F}^T}}};\\
&\cdots\cdots\cdots\cdots\cdots\cdots\cdots\cdots\cdots\cdots\cdots\cdots
\end{aligned} \tag{45}
$$

and get a new system of controls $\{g_k^\xi\}_{k=1}^\infty$, which is also a basis in $\mathcal{F}^{T,\,\xi}$ and satisfies

$$\left(C^T g_k^\xi\,,\, g_l^\xi\right)_{\mathcal{F}^T} = \delta_{kl} \qquad k,l = 1,2,\dots \tag{46}$$

by construction. The corresponding waves $\{u_k^\xi\}_{k=1}^\infty\,,\ u_k^\xi := u^{g_k^\xi}(\,\cdot\,,T) = W^T g_k^\xi$ form a basis in the subspace $\mathcal{H}^{c\xi}$, the basis turning out to be orthonormal. Indeed, we have

$$\left(u_k^\xi\,,\, u_l^\xi\right)_{\mathcal{H}^{cT}} = \langle \text{see (34)} \rangle = \left(C^T g_k^\xi\,,\, g_l^\xi\right)_{\mathcal{F}^T} = \\ \langle\ \text{see (46)} \rangle = \delta_{kl} \qquad k,l = 1,2,\dots\,.$$

We say $\{u_k^\xi\}_{k=1}^\infty$ to be a *wave basis* in $\mathcal{H}^{c\xi}$.

3.4 Truncation

As in the previous section, we keep $\xi \in (0,T)$ fixed. In the inner space $\mathcal{H}^{cT}$ introduce the operation $P^{c\xi}$

$$\left(P^{c\xi} y\right)(x) := \begin{cases} y(x)\,, & x \in \Omega^{c\xi} \\ 0\,, & x \in \Omega^{cT} \setminus \Omega^{c\xi} \end{cases}$$

that truncates functions onto the subinterval $\Omega^{c\xi} \subset \Omega^{cT}$. As is easy to see, $P^{c\xi}$ is the orthogonal projection in $\mathcal{H}^{cT}$ onto the subspace $\mathcal{H}^{c\xi}$ and, expanding over the wave basis, we can represent the truncated function in the form of the Fourier series:

$$P^{c\xi}\, y = \sum_{k=1}^{\infty} \left(y\,,\, u_k^\xi\right)_{\mathcal{H}^{cT}} u_k^\xi\,. \tag{47}$$

By the controllability (43), the function $y \in \mathcal{H}^{cT}$ can be regarded as a wave produced by a control $f = \left(W^T\right)^{-1} y$. By the same reason, the truncated function $P^{c\xi}\, y \in \mathcal{H}^{c\xi}$ is also *a wave* produced by a control $f^\xi := \left(W^T\right)^{-1} P^{c\xi} y$, the control belonging to the subspace $\mathcal{F}^{T,\,\xi}$ by (44). Thus, we have a correspondence $f \mapsto f^\xi$, which is realized by an operator

$$\mathcal{P}^\xi := \left(W^T\right)^{-1} P^{c\xi}\, W^T \tag{48}$$

acting in the outer space $\mathcal{F}^T$. In other words, the truncation $y \mapsto P^{c\xi} y$ in the inner space induces a truncation-like operation $f \mapsto \mathcal{P}^\xi f$ in the outer space (see Fig 7).

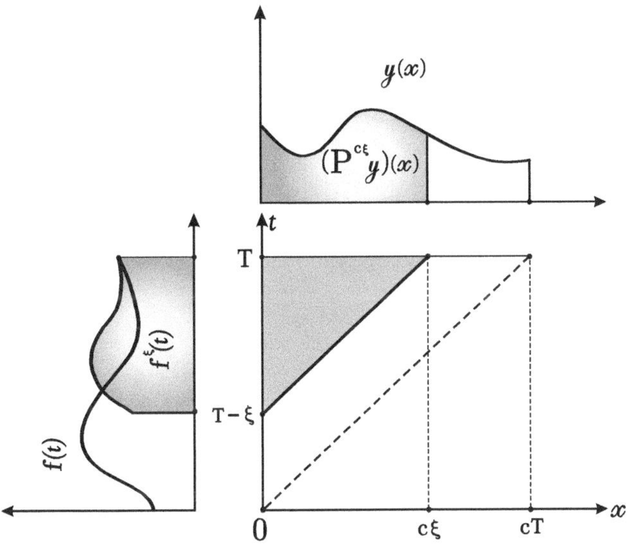

Figure 7. Truncation

An important fact is that the latter operation can be represented in the form of the expansion over the system $\{g_k^\xi\}_{k=1}^{\infty}$; namely, the relation

$$\mathcal{P}^\xi f = \sum_{k=1}^{\infty} \left(C^T f \, , \, g_k^\xi\right)_{\mathcal{F}^T} g_k^\xi \tag{49}$$

is valid. Indeed, substituting $y = W^T f$ and $u_k^\xi = W^T g_k^\xi$ in (47), we have

$$P^{c\xi} W^T f = \sum_{k=1}^{\infty} \left(W^T f \, , \, W^T g_k^\xi\right)_{\mathcal{H}^{cT}} W^T g_k^\xi = \langle \text{see (34)} \rangle$$

$$= W^T \sum_{k=1}^{\infty} \left(C^T f \, , \, g_k^\xi\right)_{\mathcal{F}^T} g_k^\xi \, .$$

Applying $\left(W^T\right)^{-1}$ and taking into account (48), we arrive at (49).

3.5 Amplitude formula

Now, let us derive a relation, which represents the values of waves through the operator $\mathcal{P}^\xi$. Let a control $f \in \mathcal{F}^T$ be continuous on $[0, T]$; by

(28) the corresponding wave $u^f(\,\cdot\,,T)$ is continuous on Ω^{cT}. The following equality holds and will play a key role in the inverse problem:

$$u^f(c\xi,T) = \left(\mathcal{P}^\xi f\right)(T-\xi+0) = \langle\text{see (49)}\rangle$$
$$= \left[\sum_{k=1}^{\infty}\left(C^T f\,,\,g_k^\xi\right)_{\mathcal{F}^T} g_k^\xi(t)\right]\Bigg|_{t=T-\xi+0}\,, \qquad 0<\xi\leq T\,. \tag{50}$$

Indeed, the truncated wave $\left(P^{c\xi}u^f(\,\cdot\,,T)\right)(x)$ vanishes for $x>c\xi$, is continuous in $\Omega^{c\xi}$, and has a jump

$$\left(P^{c\xi}u^f(\,\cdot\,,T)\right)(x)\Bigg|_{x=c\xi-0}^{x=c\xi+0} = 0\,-\,u^f(c\xi,T)\,=\,-\,u^f(c\xi,T)\,. \tag{51}$$

In the mean time, the truncated wave $P^{c\xi}u^f(\,\cdot\,,T)$ is also a *wave* produced by the "truncated" control: $P^{c\xi}u^f(\,\cdot\,,T) = u^{\mathcal{P}^\xi f}(\,\cdot\,,T)$. Hence, one can apply the jump relation (19): replacing there ξ by $T-\xi$ and taking $t=T$ in the l.h.s., we get

$$u^{\mathcal{P}^\xi f}(x,T)\Bigg|_{x=c\xi-0}^{x=c\xi+0} = -\,\mathcal{P}^\xi f(s)\Bigg|_{s=T-\xi-0}^{s=T-\xi+0} = -\,\mathcal{P}^\xi f(T-\xi+0) \tag{52}$$

because $\mathcal{P}^\xi f$ belongs to the delayed subspace $\mathcal{F}^{T,\,\xi}$ and, hence, vanishes for $0<t<T-\xi$. Thereafter, comparing (51) with (52), we easily arrive at (50).

We say the representation (50) to be the *amplitude formula* (AF). The reason is that it represents the wave through the jump amplitudes of the control, which appear as one projects the control on the subspaces $\mathcal{F}^{T,\,\xi}$ by the operators $\mathcal{P}^\xi$. The background of the AF is geometrical optics; its various versions play the role of key tool of the BC-method [7], [12].

3.6 Special BCP

In this section we deal with the problem (39) with a special r.h.s. y. Consider the Cauchy problem

$$-\,p'' + q(x)p = 0\,, \qquad x>0 \tag{53}$$
$$p\,|_{x=0} = \alpha, \quad p'\,|_{x=0} = \beta \tag{54}$$

(α and β are the constants). By standard ODE theory, such a problem has a unique smooth solution $p=p_{\alpha\beta}(x)$. In what follows we assume that α,β are fixed. Therefore we omit these subscripts and write just $p(x)$.

Consider the *special BCP*: find a control $f \in \mathcal{F}^T$ satisfying

$$u^f(\,\cdot\,,T) = p \qquad \text{in } \Omega^{cT}\,. \tag{55}$$

By (43), such a problem has a unique solution $f^T \in \mathcal{F}^T$ and, applying the AF (50) for $f = f^T$ we get

$$p(c\xi) = \left[\sum_{k=1}^{\infty} \left(C^T f^T, g_k^\xi\right)_{\mathcal{F}^T} g_k^\xi(t)\right]\Bigg|_{t=T-\xi+0}, \quad 0 < \xi \le T\,. \tag{56}$$

The remarkable fact is that, in this case, the coefficients of the series can be found explicitly.

To find the coefficients, let us take a smooth control $g \in \mathcal{F}^T$ provided $g(0) = g'(0) = g''(0) = 0$. The corresponding wave u^g satisfies (1)–(3) in the classic sense; therefore it vanishes at its forward front together with the derivatives and we have

$$u^g(cT,T) = u^g_x(cT,T) = 0\,. \tag{57}$$

This enables one to justify the following calculations:

$$\begin{aligned}
&\left(C^T f^T, g\right)_{\mathcal{F}^T} = \langle \text{see } (34)\rangle = \left(u^{f^T}(\,\cdot\,,T), u^g(\,\cdot\,,T)\right)_{\mathcal{H}^{cT}} = \\
&\langle \text{see } (55)\rangle = \int_0^{cT} p(x)\, u^g(x,T)\, \rho\, dx = \\
&\langle \text{by the zero Cauchy data (2) for the wave } u^g\rangle = \\
&\int_0^{cT} p(x) \left[\int_0^T (T-t)\, u^g_{tt}(x,t)\, dt\right] \rho\, dx = \\
&\int_0^T dt\, (T-t) \int_0^{cT} \langle p(x)\,, \rho\, u^g_{tt}(x,t)\rangle\, dx = \langle \text{see } (1)\rangle = \\
&\int_0^T dt\, (T-t) \int_0^{cT} p(x)\, \left[u^g_{xx}(x,t) - q(x) u^g(x,t)\right] dx = \langle \text{see } (57)\rangle
\end{aligned} \tag{58}$$

$$= \int_0^T dt\,(T-t)\{-p(0)u^g_x(0,t) + p'(0)u^g(0,t) +$$

$$\int_0^{cT} [p''(x) - q(x)p(x)]\; u^g(x,t)\,dx\,\} =$$

$$\langle \text{see } (53), (54)\rangle = \int_0^T (T-t)\left[-\alpha\left(R^T g\right)(t) + \beta g(t)\right]\,dt =$$

$$\left(\varkappa^T,\, -\alpha R^T g \,+\, \beta g\right)_{\mathcal{F}^T} \,=\, \left(-\left(R^T\right)^* \varkappa^T\alpha \,+\, \varkappa^T\beta\,,\, g\right)_{\mathcal{F}^T},$$

where $\varkappa^T \in \mathcal{F}^T$, $\varkappa(t) := T - t$. Thus, we have got the equality

$$\left(C^T f^T, g\right)_{\mathcal{F}^T} = \left(-\left(R^T\right)^* \varkappa^T\alpha \,+\, \varkappa^T\beta\,,\, g\right)_{\mathcal{F}^T} \tag{59}$$

for classical g's. Since such controls constitute a dense set in $\mathcal{F}^T$, the evident limit passage extends (59) to arbitrary $g \in \mathcal{F}^T$.

Taking in (59) $f = g_k^\xi$ and substituting in (56), we arrive at the following important representation:

$$p(c\xi) \,=\, \left[\sum_{k=1}^{\infty}\left(-\left(R^T\right)^* \varkappa^T\alpha \,+\, \varkappa^T\beta\,,\, g_k^\xi\right)_{\mathcal{F}^T} g_k^\xi(t)\right]\Bigg|_{t=T-\xi+0}, \quad 0 < \xi \le T\,. \tag{60}$$

The significance of this formula is that it represents the function p, which is an object of the *inner space*, through the objects of the *outer space*. We shall see that it is the fact, which enables the external observer to use (60) for solving the inverse problem.

3.7 Gelfand-Levitan-Krein equations

There exists one more way to characterize the solution of the special BCP (55) in intrinsic terms of the outer space: the control $f = f^T$ *coincides with a unique solution of the equation*

$$C^T f \,=\, -\,\alpha\left(R^T\right)^* \varkappa^T \,+\, \beta\varkappa^T\,. \tag{61}$$

In more detail, in accordance with (38) and (32), this equation takes the form

$$\sqrt{\rho}\, f(t) + \int_0^T \left[\frac{1}{2}\int_{|t-s|}^{2T-t-s} r(\eta)\, d\eta\right] f(s)\, ds =$$

$$\alpha\left[-\sqrt{\rho} + \int_t^T r(s-t)\,(T-s)\, ds\right] + \beta\,(T-t)\,, \qquad 0<t<T\,. \tag{62}$$

The relation (61) follows from (59) just by arbitrariness of g. Applying the jump relation (20) for $\xi = T$, we obtain

$$f^T(+0) = u^{f^T}(cT-0, T) = p(cT)\,. \tag{63}$$

For solving the inverse problem, not a single equation (61) will be enlisted but a family of such equations corresponding to the family of the special BCP's

$$u^f(\,\cdot\,,\xi) = p \qquad \text{in } \Omega^{c\xi}\,, \quad (0<\xi\le T). \tag{64}$$

Just replacing T by ξ in the previous considerations, we conclude that the solution $f = f^\xi \in \mathcal{F}^\xi$ of (64) coincides with a unique solution of the equation

$$C^\xi f = -\alpha\left(R^\xi\right)^* \varkappa^\xi + \beta\varkappa^\xi \tag{65}$$

or, equivalently,

$$\sqrt{\rho}\, f(t) + \int_0^\xi \left[\frac{1}{2}\int_{|t-s|}^{2\xi-t-s} r(\eta)\, d\eta\right] f(s)\, ds -$$

$$\alpha\left[-\sqrt{\rho} + \int_t^\xi r(s-t)\,(\xi-s)\, ds\right] + \beta\,(\xi-t)\,, \qquad 0<t<\xi\,. \tag{66}$$

Also, in perfect analogy to (63), the relations

$$f^\xi(+0) = p(c\xi)\,, \qquad 0<\xi\le T \tag{67}$$

are valid.

For each ξ, the equation (65) is a 2-order Fredholm integral equation. Since W^ξ is a boundedly invertible operator (see sec 3.2), the same holds for the connecting operator $C^\xi = \left(W^\xi\right)^* W^\xi$, which turns out to be a positive definite isomorphism in $\mathcal{F}^\xi$. Hence, each equation is uniquely and stably solvable.

The family of the equations (66) plays the key role in solving the inverse problem. We refer to them as the Gelfand-Levitan-Krein equations (GLK). The reason is that the classical Gelfand-Levitan equations, which for the first time were derived for solving the spectral inverse problem for the Sturm–Liouville operator (see [19]), can be obtained from (66) by differentiation and simple change of variables. In the same way, dealing with the special BCP for the inhomogeneous string with a variable density [7], one can derive the relevant analog of (66) and then reduce it to the classical M.Krein equation [23]–[25], as is done in pioneer works by Sondhi and Gopinath [20], [21] and Blagovestchenskii [17], [18] (see [8] for detail). So, in the framework of the Boundary Control method, the classical inverse problem equations are interpreted as *the equations in the outer space of the system α^T, which provide the solutions of the relevant special BCP's.* To the best of our knowledge, for the first time, such a look at the Krein equation was proposed in [20]. It is also relevant in multidimensional problems [3].

4 Inverse Problem

4.1 Statement

The setup of the dynamical inverse problem is motivated by the locality principle or, more exactly, by the local character of dependence of the response operator on the potential. Recall that the operator R^{2T} associated with the extended problem 1 is determined by $q\,|_{\Omega^{cT}}$ (see the remark below (33)). This property is of transparent physical meaning. Namely, the response of the system α^T (i.e., the force $u^f_x(0,t)$ measured at the endpoint of the string) on the action of a control f is formed by the waves, which are reflected from inhomogeneities of the string and return back to the endpoint $x=0$ (see the arrows going to the left on Fig 8). Since the wave propagation speed is equal to c, the waves reflected from the depths $x > cT$ return to the endpoint *later* than $t = 2T$; they are not recorded by the external observer measuring the values $u^f_x(0,t)$ for times $0 \le t \le 2T$. Therefore, the operator R^{2T}, which corresponds to those measurements, contains certain information on $q\,|_{\Omega^{cT}}$ but "knows nothing" about $q|_{x>cT}$. Taking into account this fact, the relevant statement of the dynamical inverse problem has to be as follows: **given operator R^{2T}, recover the potential $q\,|_{\Omega^{cT}}$.**

[7] Such a string is described by the equation $\rho(x)u_{tt} - u_{xx} = 0$

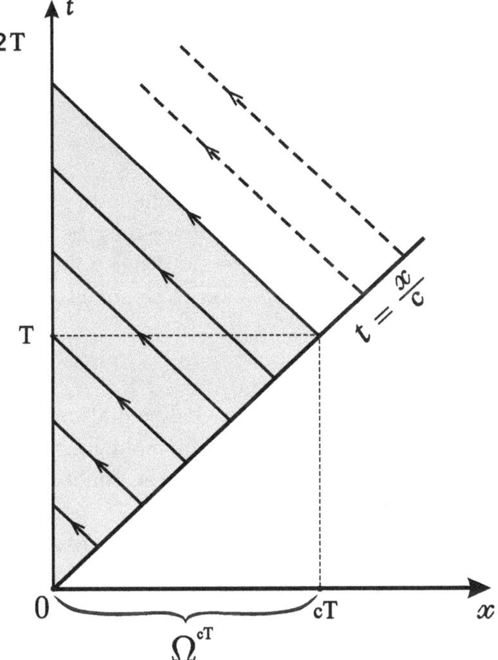

Figure 8. Incident and reflected waves

Note in addition that this statement agrees closely with very general principles of the system theory [22]. By one of them, the input$\mapsto$output map of a linear system (here the response operator R^{2T}) determines not the whole system but its controllable part (here the interval Ω^{cT}) only.

4.2 Solving inverse problem

The potential $q\mid_{\Omega^{cT}}$ can be recovered by means of the following procedure:

Step 1 Determine the constant $\sqrt{\rho}$ and the reply function $r\mid_{(0,2T)}$ from (33). Find the operator C^T by (38). Compose $\left(R^T\right)^*$ by (32).

Step 2 Fix $\xi \in (0,T]$. In the subspace $\mathcal{F}^{T,\xi} \subset \mathcal{F}^T$, construct a C^T-orthogonal basis of controls $\{g_k^\xi\}_{k=1}^\infty$ (see (45)).

Step 3 Choose two constants α, β provided $\alpha^2+\beta^2 \neq 0$ and find the value $p(c\xi)$ by the use of the relation (60).

Step 4 Varying $\xi \in (0,T]$, recover the function p in Ω^{cT}.

Step 5 Determine the potential from the equation (53):

$$q(x) = \frac{p''(x)}{p(x)}, \qquad x \in \Omega^{cT}. \tag{68}$$

Surely, using (68) one has to take care of the case $p(x) = 0$. However, as is well known, on any finite interval Ω^{cT} there may be only a finite number of zeros of p. Therefore, if q is determined in other points it can be extended to the zeros by continuity. Another option to remove the zero of p at a given $x = x_0$ is to choose another α, β.

The inverse problem is solved.

One more way is to make use of the GLK-equations. One can compose the family of the equations (66), find their solutions $f = f^\xi$, determine the values $p(c\xi)$ by (67), and hence recover the function p in Ω^{cT}. Thereafter, the potential is determined by (68). In [15], such a scheme is realized numerically for the string of variable density governed by the equation $\rho u_{tt} - u_{xx} = 0$.

So, the dynamical inverse problem stated in sec 4.1 is solved. In such a clear and natural form it was set up by A.S.Blagovestchenskii, who elaborated the so-called *local approach* solving the problem for the inhomogeneous string and providing the characteristic conditions for its solvability [17]. Independently and almost simultaneously, in physical literature there was proposed another approach belonging to B.Gopinath and M.M.Sondhi [20], [21], which also solves the inverse problem in the above-mentioned setup, i.e., preserves the *locality of determination* of the string parameters. Both of the approaches derive M.Krein's equation and apply it for solving the inverse problem, the derivation being based upon a purely local (hyperbolic) technique that allows not to invoke irrelevant spectral devices (like the Fourier transform w.r.t. time, etc). A peculiarity of the Gopinath–Sondhi approach is its straightforward relation to the BCP.

4.3 Visualization of waves

Here we realize the option mentioned at the end of sec 3.1 and show how one can make the waves on the string to be visible for the external observer. The proposed procedure is a core of the BC-method: it is relevant to a wide class of inverse problems including *multidimensional* ones [4], [7].

Let the observer choose a smooth control $f \in \mathcal{F}^T$, so that the corresponding (invisible) wave $u^f(\,\cdot\,, T)$ is also smooth in Ω^{cT}.

Step 1 Fix $\xi \in (0,T]$ and construct the system of controls $\{g_k^\xi\}_{k=1}^\infty \subset \mathcal{F}^{T,\xi}$ that produces the (invisible) wave basis by (45).

Step 2 Determine the operator $\mathcal{P}^\xi$ by (49) and then find the value $u^f(c\xi,T)$ by (50).

Step 3 Varying $\xi \in (0,T]$ and repeating the same trick, recover the wave $u^f(x,T)$ in Ω^{cT}.

So, we succeeded in visualizing the wave $u^f(\,\cdot\,,T)$! Moreover, replacing ξ by t in the delay relation (29), we can recover the wave process in a whole:

$$u^f(c\xi,t) = \left(\mathcal{P}^\xi D^{T,t} f\right)(T-\xi+0)\,, \qquad 0<\xi,t\le T\,, \tag{69}$$

what is much more than just to recover $q\,|_{\Omega^{cT}}$. Indeed, as soon as the evolution of the wave is seen, the potential can be determined, e.g., from the wave equation (1).

By the way, the representation (69) is quite available for constructing numerical algorithms. The only thing, which is needed for the use of (60), is to prepare the system of controls $\{g_k^\xi\}$. To this end, one can take a rich enough initial system of controls $\{f_j^\xi\}_{j=1}^N \subset \mathcal{F}^{T,\xi}$, for instance,

$$f_j^\xi(t) = \begin{cases} \frac{N}{\xi}\,, & (T-\xi)+(j-1)\frac{\xi}{N} \le t \le (T-\xi)+j\frac{\xi}{N} \\ 0\,, & \text{for other } \; t\in[0,T] \end{cases} \tag{70}$$

and then apply the Schmidt process (45). As is well known, such an orthogonalization is equivalent to inverting the Gram matrix $\{(C^T f_i^\xi, f_j^\xi)_{\mathcal{F}^T}\}_{i,j=1}^N$. Since the connecting operator is boundedly invertible in $\mathcal{F}^T$, this matrix turns out to be well posed uniformly w.r.t. N that provides the process to be stable. However, to recover the potential by the use of (68) one needs to invoke the numerical differentiation that complicates the calculations but can be realized by standard regularizing procedures. Regarding the use of the GLK-equations (66), note that, from the computational viewpoint, to solve them and to invert the Gram matrix of the large size is, roughly speaking, one and the same [15].

APPENDIX

I. Structure of fundamental solution The sequence of functions and distributions $\{\theta^j(t)\}_{j=-\infty}^{\infty}$, $-\infty < t < \infty$,

$$\theta^0(t) := \begin{cases} 0, & t<0 \\ 1, & t \geq 0 \end{cases}; \qquad \theta^j(t) = \int_{-\infty}^{t} \theta^{j-1}(s)\,ds\,,$$

$$j = \ldots, -2, -1, 0, 1, 2, \ldots \tag{71}$$

is said to be a *smoothness scale*: the bigger j, the smoother θ^j and

$$\theta^{j-1}(t) = \frac{d\theta^j}{dt}(t)\,, \qquad -\infty < t < \infty$$

holds. In particular,

$$\ldots\,,\ \theta^{-3} = \delta''(t)\,,\ \theta^{-2} = \delta'(t)\,,\ \theta^{-1} = \delta(t)\,,\ \ldots\,,\ \theta^j = \frac{t^j}{j!}\,\theta^0(t)\,.$$

Note that all elements of the scale vanish as $t < 0$.

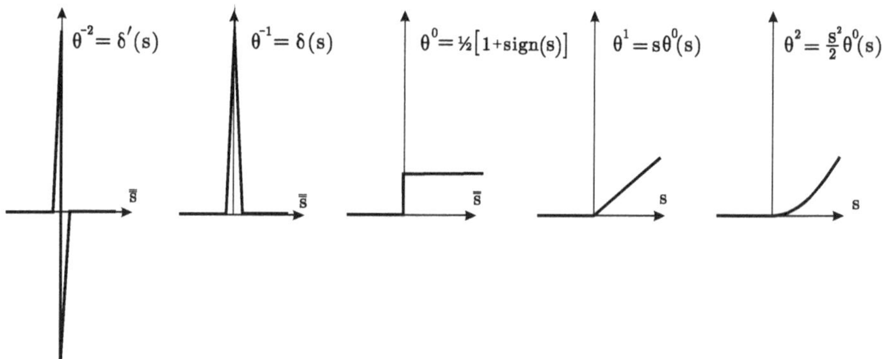

Figure 9. Smoothness scale

Let us look for the fundamental solution in the form of a formal series

$$u^\delta(x,t) = \sum_{j=-1}^{\infty} a^j(x)\,\theta^j\left(t - \frac{x}{c}\right) \tag{72}$$

with unknown functions a_j, which can be referred to as the Taylor expansion of u^δ near (from the left of) its forward front $t = \frac{x}{c}$. The reason to begin the series with $j = -1$ is that in the unperturbed case one has $\breve{u}^\delta(x,t) = \delta\left(t - \frac{x}{c}\right) = \theta^{-1}\left(t - \frac{x}{c}\right)$. Differentiating and combining the similar terms, we easily get

$$\frac{1}{c^2}\, u^\delta_{tt} - u^\delta_{xx} + qu^\delta = \left[\frac{2}{c}\, a'_{-1}(x)\right] \theta^{-2}\left(t - \frac{x}{c}\right) +$$
$$\sum_{j=-1}^{\infty}\left[\frac{2}{c}\, a'_{j+1}(x) - a''_j(x) + q(x)a_j(x)\right] \theta^j\left(t - \frac{x}{c}\right) = 0\,.$$

By the independence of the different order singularities, we get the recurrent system of the ODEs

$$\frac{2}{c}\, a'_{-1} = 0\,, \qquad \frac{2}{c}\, a'_j - a''_{j-1} + qa_{j-1} = 0\,, \qquad j = 0\,,1\,,\,\ldots\,,$$

which is often called the *transport equations* and can be integrated one by one. In the mean time, the condition $u^\delta(0,t) = \delta(t), t \geq 0$ implies the initial conditions

$$a'_{-1}(0) = 1\,, \qquad a'_j(0) = 0\,, \qquad j = 0\,,1\,,\,\ldots\,.$$

The integration yields

$$a'_{-1} - 1\,,\ a'_0(x) - -\frac{c}{2}\int_0^x q(s)\,ds\,,\ \ldots\ .$$

Summarizing, we arrive at the well-known representation

$$u^\delta(x,t) = \delta\left(t - \frac{x}{c}\right) - \left[\frac{c}{2}\int_0^x q(s)\,ds\right]\theta^0\left(t - \frac{x}{c}\right) + \ \ldots$$

that describes the behavior of the fundamental solution near its forward front $t = \frac{x}{c}$.

The above-presented derivation is often referred to as a dynamical version of the *WKB-method*, whereas (72) is called the *progressive wave expansion* [8]. The meaning of such a representation can be commented on as follows. Even though the shape of a wave can be complicated, the evolution of its singularities (δ-terms, jumps, etc) is ruled by rather simple lows

[8] The terms *ray method* and *ray expansion* are also in use.

(geometrical optics formulas). This fact is also valid for a wide class of hyperbolic problems including multidimensional ones.

II. Proof of (37), (38) Choose smooth controls $f, g \in \mathcal{F}^T$ such that f vanishes near $t = 0$ and $t = T$, whereas g vanishes near $t = 0$. Denote $f_- := S^T f \in \mathcal{F}^{2T}$; let u^g be the solution of (1)–(3) and u^{f_-} the solution of (21)–(23) such that $u|_{x=0} = f_-$. Note that both of these solutions are classical. Also, by the choice of g and the property (17), for every $t > 0$ the solution $u^g(x,,t)$ vanishes in a neighborhood of $x = ct$.

Introduce Blagovestchenscii's function

$$b(s,t) := \int_0^\infty u^{f_-}(x,s)\, u^g(x,t)\, \rho\, dx, \qquad (s,t) \in B^{2T}$$

in the triangular $B^{2T} := \{(s,t) \,|\, 0 \le t \le T,\ t \le s \le 2T - t\}$. Function b satisfies the relations

$$\begin{aligned}
& b_{tt}(s,t) - b_{ss}(s,t) = \\
& \int_0^\infty \left[u^{f_-}(x,s)\, u^g_{tt}(x,t) - u^{f_-}_{ss}(x,s)\, u^g(x,t)\right] \rho\, dx = \langle \text{see } (1) \rangle = \\
& \int_0^\infty [u^{f_-}(x,s)\ (u^g_{xx}(x,t) - q(x)u^g(x,t)) - \\
& \left(u^{f_-}_{xx}(x,s) - q(x)u^{f_-}(x,s)\right)\ u^g(x,t)]\, dx = \\
& \int_0^\infty \left[u^{f_-}(x,s)\, u^g_{xx} - u^{f_-}_{xx}(x,s)\, u^g(x,t)\right]\, dx = \\
& - \left[u^{f_-}(0,s)\, u^g_x(0,t) - u^{f_-}_x(x,s)\, u^g(0,t)\right] = \\
& - f_-(s)\ \left(R^T g\right)(t) + \left(R^{2T} f_-\right)(s)\, g(t) =: F(s,t)
\end{aligned}$$

(in course of integration by parts, the terms at $x = \infty$ vanish by the above mentioned properties of u^g). Hence, with regard to the zero Cauchy data (2) for u^g, we arrive at the problem

$$\begin{aligned}
& b_{tt} - b_{ss} = F \quad \text{in } B^{2T} \\
& b\big|_{t=0} = b_t\big|_{t=0} = 0, \qquad 0 \le s \le 2T.
\end{aligned}$$

Applying the D'Alembert formula, we have

$$b(s,t) = \frac{1}{2}\int\limits_0^t d\eta \int\limits_{s-t+\eta}^{s+t-\eta} F(\xi,\eta)\, d\xi, \qquad (s,t) \in B^{2T}.$$

Putting here $s = t = T$, we get

$$\begin{aligned}
&b(T,T) = \\
&\frac{1}{2}\int\limits_0^T d\eta \int\limits_\eta^{2T-\eta} \left[-f_-(\xi)\left(R^T g\right)(\eta) + \left(R^{2T} f_-\right)(\xi)\, g(\eta)\right] d\eta = \\
&-\frac{1}{2}\int\limits_0^T d\eta\, \left(R^T g\right)(\eta) \int\limits_\eta^{2T-\eta} f_-(\xi)\, d\xi + \\
&+\int\limits_0^T d\eta\; g(\eta)\, \frac{1}{2}\left[\int\limits_0^{2T-\eta} \left(R^{2T} f_-\right)(\xi)\, d\xi - \int\limits_0^\eta \left(R^{2T} f_-\right)(\xi)\, d\xi\right].
\end{aligned}$$

By oddness of f_-, we have

$$\int\limits_\eta^{2T-\eta} f_-(\xi)\, d\xi \equiv 0, \quad 0 < \eta < T.$$

Going on the calculation and taking into account the representation

$$\begin{aligned}
&\frac{1}{2}\left[\int\limits_0^{2T-\eta} \left(R^{2T} f_-\right)(\xi)\, d\xi - \int\limits_0^\eta \left(R^{2T} f_-\right)(\xi)\, d\xi\right] = \langle \text{see } (35)\rangle = \\
&-\frac{1}{2}\left(\left(S^T\right)^* J^{2T} R^{2T} S_-^T f\right)(\eta) = \langle \text{see } (36)\rangle = \\
&-\frac{1}{2}\left(\left(S^T\right)^* R^{2T} J^{2T} S_-^T f\right)(\eta), \qquad 0 < \eta < T
\end{aligned}$$

one obtains

$$b(T,T) = \left(-\frac{1}{2}\left(S^T\right)^* R^{2T} J^{2T} S^T f,\, g\right)_{\mathcal{F}^T}.$$

On the other hand, the definition of b and the evident equality $u^{f_-}(\,\cdot\,,T) = u^f(\,\cdot\,,T)$ imply

$$b(T,T) = \int_0^\infty u^{f_-}(x,T)\, u^g(x,T)\, \rho\, dx = \big(u^f(\cdot,T), u^g(\cdot,T)\big)_{\mathcal{H}^{cT}} =$$
$$\langle \text{ see } (34)\rangle = \big(C^T f, g\big)_{\mathcal{F}^T}\,.$$

Compare two obtained expressions for $b(T,T)$. Taking into account the arbitrariness of the chosen f, g and density of such controls in $\mathcal{F}^T$, we justify the representation (37). Substituting (33) in it, one can easily derive (38).

An important opportunity to express the inner product of waves via the response operator was found out by A.S. Blagovestchenskii.

PART II: 2-CHANNEL DYNAMICAL SYSTEM (BEAM)

5 Forward problem

5.1 Statement

Here we deal with an initial boundary value problem (*problem* 1) for the system

$$\rho u_{tt} - u_{xx} + A u_x + B u = 0, \qquad x > 0,\ \ 0 < t < T \tag{73}$$
$$u|_{t=0} = u_t|_{t=0} = 0, \qquad x \geq 0 \tag{74}$$
$$u|_{x=0} = f, \qquad 0 \leq t \leq T\,, \tag{75}$$

where

$$\rho = \begin{pmatrix} \rho_1 & 0 \\ 0 & \rho_2 \end{pmatrix},\ A = \begin{pmatrix} a_{11}(x) & a_{12}(x) \\ a_{21}(x) & a_{22}(x) \end{pmatrix},\ B = \begin{pmatrix} b_{11}(x) & b_{12}(x) \\ b_{21}(x) & b_{22}(x) \end{pmatrix}$$

are 2×2- matrices with smooth (continuously differentiable) entries, whereas $\rho_{1,2}$ are the constants provided $0 < \rho_1 < \rho_2$ [9]; $f = \begin{pmatrix} f_1(t) \\ f_2(t) \end{pmatrix}$ is a boundary control; $u = u^f(x,t) = \begin{pmatrix} u_1^f(x,t) \\ u_2^f(x,t) \end{pmatrix}$ is a solution (wave).

[9] The case $0 < \rho_1 = \rho_2$ will be mentioned later in Comments.

Also we impose the following *self-adjointness condition* on the coefficients:

$$A^\sharp(x) = -A(x), \quad \frac{dA}{dx}(x) = B(x) - B^\sharp(x), \qquad x \geq 0, \tag{76}$$

where $(\dots)^\sharp$ is the matrix conjugation; in more detail,

$$A(x) = \begin{pmatrix} 0 & -a(x) \\ a(x) & 0 \end{pmatrix}, \; b_{21}(x) - b_{12}(x) = a'(x), \qquad x \geq 0.$$

Such conditions come from physics: they provide a relevant energy conservation low on the beam [10].

In parallel we consider the *unperturbed problem* $\breve{1}$

$$\rho u_{tt} - u_{xx} = 0, \qquad x > 0, \; 0 < t < T \tag{77}$$

$$u|_{t=0} = u_t|_{t=0} = 0, \qquad x \geq 0 \tag{78}$$

$$u|_{x=0} = f, \qquad 0 \leq t \leq T, \tag{79}$$

and denote by $\breve{u}^f$ its solution, which can be found in explicit form:

$$\breve{u}^f(x,t) = \begin{pmatrix} f_1(t - \frac{x}{c_1}) \\ f_2(t - \frac{x}{c_2}) \end{pmatrix} \tag{80}$$

(recall the Convention of sec 1.2!), where $c_i := \frac{1}{\sqrt{\rho_i}}$, $c_1 > c_2$. Such a solution describes a wave process in two-channel system, in which two different wave modes $\begin{pmatrix} u_1^f(x,t) \\ 0 \end{pmatrix}$ and $\begin{pmatrix} 0 \\ u_2^f(x,t) \end{pmatrix}$ propagate in the first (*fast*) and second (*slow*) channels with the velocities c_1 and c_2 respectively, the modes propagating independently (not interacting with each other).

In contrast to this, in the perturbed system (73)–(75) with non-diagonal A and/or B, the modes do interact and we shall see that such an interaction provides interesting effects. A classical example of a 2-channel system with interacting modes is a Timoshenko beam in elasticity theory. One more example comes from electrical engineering: a twin-core cable with interacting cores.

[10] However, all results of general character (sec 5.2 – 5.5) are valid for arbitrary smooth coefficients A, B.

5.2 Integral equation and generalized solutions

The problem 1 can be reduced to a system of integral equations by means of the same trick that was used in sec 2.2. Namely, denoting $Qu := Au_x + Bu$, representing $u^f = \breve{u}^f + w$, and substituting to (73)–(75) with a new unknown vector-function $w = \begin{pmatrix} w_1^f(x,t) \\ w_2^f(x,t) \end{pmatrix}$, we get the problem

$$\begin{aligned} &\rho w_{tt} - w_{xx} = -Qw - Q\breve{u}^f, && x > 0,\ 0 < t < T \\ &u|_{t=0} = u_t|_{t=0} = 0, && x \geq 0 \\ &u|_{x=0} = f, && 0 \leq t \leq T\,. \end{aligned}$$

Applying the D'Alembert formula by components, we arrive at the equation

$$w + Mw = -M\breve{u}^f \tag{81}$$

with the matrix integral operator M, which acts by the rule

$$(Mw)(x,t) = \begin{pmatrix} \frac{1}{2c_1} \int\int_{K_{c_1}(x,t)} (Qw)_1(\xi,\eta)\, d\xi\, d\eta \\ \frac{1}{2c_2} \int\int_{K_{c_2}(x,t)} (Qw)_2(\xi,\eta)\, d\xi\, d\eta \end{pmatrix}, \qquad x \geq 0\,.$$

and $K_{c_i}(x,t)$ are the trapeziums bounded by the characteristic lines $t \pm \frac{x}{c_i} =$ const.

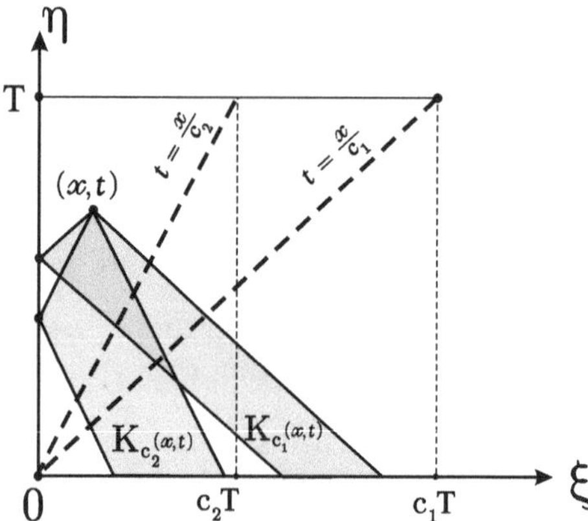

Figure 10. Domains $K_{c_i}(x,t)$

The equation (81) is a second kind Volterra- type equation which can be analyzed by iteration method. The analysis provides the following results (see [16], [31] for detail):

- If the control f is a twice continuously differentiable (vector-) function provided $f(0) = f'(0) = f''(0) = 0$, then one can check that the solution w^f of (81) is also twice differentiable w.r.t. x, t, whereas the function $u^f = \breve{u}^f + w^f$ turns out to be a unique *classical solution* of the problem (73)– (75).
- As is easy to see, for $f \in L_2\left((0,T);\mathbb{R}^2\right)$[11], the r.h.s. of the equation (81) is a continuous function of the variables x, t, the function vanishing for $t < \frac{x}{c_1}$. Simple analysis shows that the equation is uniquely solvable in this class of functions. In this case, the function $u^f := \breve{u}^f + w^f$ is regarded as a *generalized solution* of the problem 1. So, just by the accepted definition such a solution does always exist and is unique.

5.3 Fundamental solution

For the problem 1, the relevant analog of the scalar fundamental solution u^δ is the solution of the matrix problem

$$\rho U_{tt} - U_{xx} + AU_x + BU = 0, \qquad x > 0,\ 0 < t < T \tag{82}$$

$$U|_{t=0} = U_t|_{t=0} = 0, \qquad x \geq 0 \tag{83}$$

$$U_{x=0} = \delta(t)\begin{pmatrix}1 & 0\\ 0 & 1\end{pmatrix}, \qquad 0 \leq t \leq T, \tag{84}$$

where δ is the Dirac delta-function, $U = U(x,t)$ is a 2×2-matrix-function. In the unperturbed case $A = B = 0$ holds and one has

$$\breve{U}(x,t) = \begin{pmatrix}\delta(t-\frac{x}{c_1}) & 0\\ 0 & \delta(t-\frac{x}{c_2})\end{pmatrix}. \tag{85}$$

In the perturbed case, representing

$$U = \breve{U} + W, \tag{86}$$

one can reduce the problem to a relevant version of the matrix integral equation (81) for the regular part W and analyze it by the iteration method. Omitting the details, let us present some results.

1. The solution $W(x,t)$ is a piece-wise smooth matrix-function, whereas possible breaks of smoothness (jumps and/or jumps of derivatives) can occur on the characteristic lines $t = \frac{x}{c_i}$ only. The solution is supported above the fast characteristic:

$$W(x,t)\Big|_{t<\frac{x}{c_1}} = 0 \tag{87}$$

[11] It is the class of vector-functions satisfying $\int_0^T \left[(f_1(t))^2 + (f_2(t))^2\right] dt < \infty$.

and satisfies

$$W(0,t) = 0\,, \qquad t > 0\,. \tag{88}$$

2. To describe a structure of the fundamental solution let us note that the columns of the matrix U coincide with the solutions of the problem 1 corresponding to the controls $\begin{pmatrix}\delta(t)\\0\end{pmatrix}$ and $\begin{pmatrix}0\\\delta(t)\end{pmatrix}$:

$$U(x,t) = \begin{pmatrix} u_1^{\begin{pmatrix}\delta\\0\end{pmatrix}}(x,t) & u_1^{\begin{pmatrix}0\\\delta\end{pmatrix}}(x,t) \\ u_2^{\begin{pmatrix}\delta\\0\end{pmatrix}}(x,t) & u_2^{\begin{pmatrix}0\\\delta\end{pmatrix}}(x,t) \end{pmatrix} = \langle \text{see (85)} \rangle =$$

$$\begin{pmatrix} \delta(t-\frac{x}{c_1}) + W_{11}(x,t) & W_{12}(x,t) \\ W_{21}(x,t) & \delta(t-\frac{x}{c_2}) + W_{22}(x,t) \end{pmatrix}\,. \tag{89}$$

By (87), the fundamental solution vanishes under the fast characteristic $t = \frac{x}{c_1}$; typical behavior of its components is illustrated on Fig 11 and commented on below:

- **Fig 11a** The control $\begin{pmatrix}\delta\\0\end{pmatrix}$ injects the singularity into the first (fast) channel. The singularity of the form $\delta(t-\frac{x}{c_1})$ propagates in this channel with the velocity c_1 and interacts with inhomogeneities of the beam. As result of interaction, a tail appears.
- **Fig 11b** Owing to interaction between channels, the fast component $u_1^{\begin{pmatrix}\delta\\0\end{pmatrix}}$ induces a wave into the second (slow) channel, the wave propagating with "anomalous" fast velocity c_1. The part of the component $u_2^{\begin{pmatrix}\delta\\0\end{pmatrix}}$, which appears as result of the inter-channel connection, is called a *precursor* [12] (shaded on Fig 11b) Also note that in the case of $A = 0$ the second component turns out to be smoother: it is continuous (has no jumps) but may have jumps of derivatives at the characteristics $t = \frac{x}{c_i}$.
- **Fig 11c** The control $\begin{pmatrix}0\\\delta\end{pmatrix}$ injects the singularity into the second (slow) channel. The singularity $\delta(t-\frac{x}{c_2})$ propagates in this channel with the velocity c_2. Inhomogeneities of the beam yield appearance of the tail

[12] The term *head wave* is also in use.

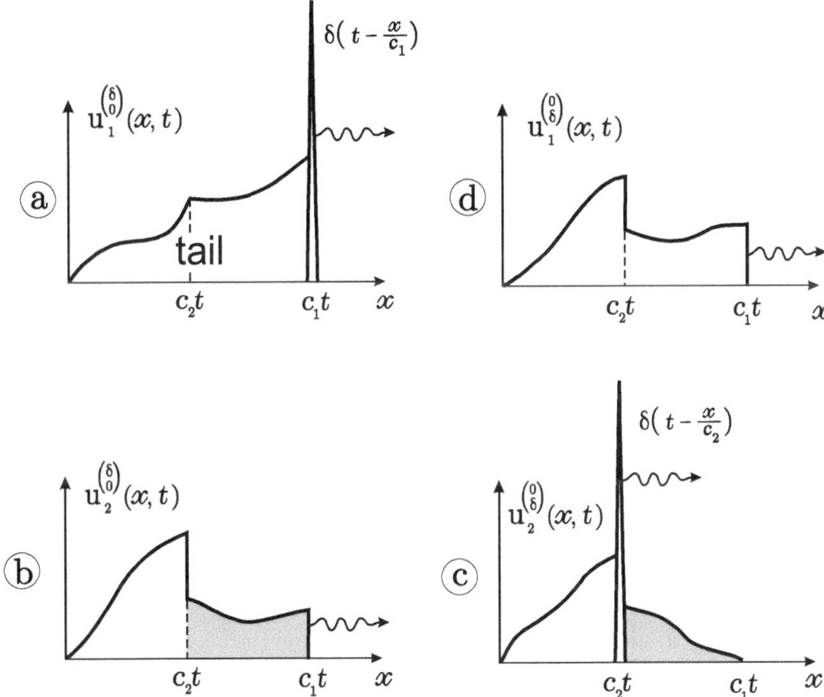

Figure 11. Fundamental matrix solution

behind the leading singularity. Since the channels do interact, the slow mode induces a wave process in the fast channel, which propagates with the velocity c_1 and, hence, leaves behind the slow singularity. In the mean time, such a process excites the waves in the slow channel, what implies the appearance of the precursor (shaded) propagating with anomalous fast velocity.

- **Fig 11d** The fast component of the wave $u^{\begin{pmatrix}0\\ \delta\end{pmatrix}}$ appears as result of the inter-channel connection.

  In the case $A = 0$ the interaction between channels is weaker and the first component turns out to be smoother: it is continuous (has no jumps) but may have jumps of derivatives at the characteristics $t = \frac{x}{c_i}$.

It is worthwhile to note that the leading singular part of the fundamental solution is the same in the perturbed and unperturbed case (see (89)).

Also some jumps (as well as jumps of derivatives) shown on Fig 11, can occasionally vanish that depends on the behavior of the coefficients A, B.

As we see, the interaction between channels renders the picture of waves on the beam much more complicated than one on the string. The technique that provides the picture shown on Fig 11 and computes the jumps of the regular part of the fundamental solution is in principle the same as the technique developed for the scalar case in Appendix I. Namely, by analogy to (72) one can seek for U in the form of the expansion by smoothness

$$U(x,t) = \begin{pmatrix} \delta(t-\frac{x}{c_1}) & 0 \\ 0 & \delta(t-\frac{x}{c_2}) \end{pmatrix} + \sum_{j=0}^{\infty}\left[a_j(x)\,\theta^j\left(t-\frac{x}{c_1}\right)+b_j(x)\,\theta^j\left(t-\frac{x}{c_2}\right)\right] \tag{90}$$

with (unknown) matrices a_j, b_j, then substitute (90) in (82) – (84) and determine the matrices by means of the separation of singularities. However, since the calculations are rather cumbersome, we omit them and recommend the reader to fill this gap as an exercise. As example, the first coefficients of the series look as follows:

$$a_0(x) = \begin{pmatrix} -\frac{1}{2\sqrt{\rho_1}}\int\limits_0^x\left[\frac{\rho_1}{\rho_2-\rho_1}\,a^2(s)+b_{11}(s)\right]ds & -\frac{\sqrt{\rho_2}}{\rho_2-\rho_1}\,a(0) \\ \frac{\sqrt{\rho_1}}{\rho_2-\rho_1}\,a(x) & 0 \end{pmatrix},$$

$$b_0(x) = \begin{pmatrix} 0 & \frac{\sqrt{\rho_2}}{\rho_2-\rho_1}\,a(x) \\ -\frac{\sqrt{\rho_1}}{\rho_2-\rho_1}\,a(0) & \frac{1}{2\sqrt{\rho_2}}\int\limits_0^x\left[\frac{\rho_2}{\rho_2-\rho_1}\,a^2(s)-b_{22}(s)\right]ds \end{pmatrix}.$$

Recall that $a(x)$ is the matrix element of the coefficient $A(x)$: see (76).

5.4 Properties of waves

Since the problem 1 is linear and the coefficients A, B are time-independent, the superposition principle does hold: by the Duhamel formula, for a control

$f \in L_2\left((0,T);\mathbb{R}^2\right)$ one has

$$u^f(x,t) = U(x,t) * f(t) = \langle \text{see } (89) \rangle =$$

$$\begin{pmatrix} f_1(t-\frac{x}{c_1}) \\ f_2(t-\frac{x}{c_2}) \end{pmatrix} + \int_0^t W(x,t-s)\, f(s)\, ds = \langle \text{see } (87) \rangle =$$

$$\begin{pmatrix} f_1(t-\frac{x}{c_1}) \\ f_2(t-\frac{x}{c_2}) \end{pmatrix} + \int_0^{t-\frac{x}{c_1}} \begin{pmatrix} W_{11}(x,t-s) & W_{12}(x,t-s) \\ W_{21}(x,t-s) & W_{22}(x,t-s) \end{pmatrix} \begin{pmatrix} f_1(s) \\ f_2(s) \end{pmatrix} ds \qquad (91)$$

Such a representation easily implies the following properties of the solution u^f:

- The relation
$$u^f \,|_{t<\frac{x}{c_1}} = 0, \qquad x \geq 0,\ t \geq 0\,, \qquad (92)$$
which is interpreted as the finiteness of the wave propagation speed on the beam, holds.
- For $f \in L_2\left((0,T);\mathbb{R}^2\right)$ and $\tau \in [0,T]$, define a delayed control $f_\tau \in L_2\left((0,T);\mathbb{R}^2\right)$ by
$$f_\tau(t) := f(t-\tau)\,, \qquad 0 \leq t \leq T\,,$$
where τ is the value of delay (recall the Convention!). Since the coefficients A, B do not depend on t, the delay of the control implies the same delay of the wave: the relation
$$u^{f_\tau}(x,t) = u^f(x,t-\tau) \qquad (93)$$
is valid.
- Let a control f be a piece-wise continuous vector-function, which has a jump at $t=\xi$. By the properties of the regular part of the fundamental solution W, the integral term in (91) depends on x,t continuously. Therefore, $u^f(x,t)$ and the term $\begin{pmatrix} f_1(t-\frac{x}{c_1}) \\ f_2(t-\frac{x}{c_2}) \end{pmatrix}$ have one and the same jumps at the characteristic lines $t = \frac{x}{c_i} + \xi$, the values of the jumps being connected (by components) as

$$u_i^f(x,t)\Big|_{x=c_i(t-\xi)-0}^{x=c_i(t-\xi)+0} = -\,f_i(s)\Big|_{s=\xi-0}^{s=\xi+0}\,, \qquad t>0\,, \quad i=1,2\,. \qquad (94)$$

So, the jumps propagate along the fast and slow characteristics see Fig 12a.

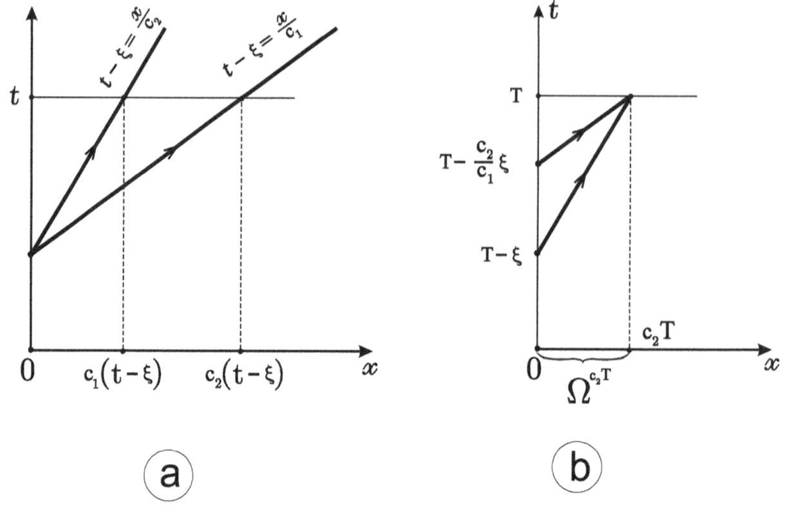

Figure 12. Propagation of jumps

For what follows it is convenient to write (94) in the following equivalent form. Assume that at the final moment $t = T$ the wave $u^f(\cdot, T)$ turns out to be a piece-wise continuous, with a jump at the point $x = c_2\xi$. As is easily seen from (91), such a case is possible only if the control f has a jump in both components, the values of jumps being connected through the relations

$$u_1^f(x,T)\Big|_{x=c_2\xi-0}^{x=c_2\xi+0} = f_1(s)\Big|_{s=T-\frac{c_2}{c_1}\xi+0}^{s=T-\frac{c_2}{c_1}\xi-0}, \quad u_2^f(x,T)\Big|_{x=c_2\xi-0}^{x=c_2\xi+0} = f_2(s)\Big|_{s=T-\xi+0}^{s=T-\xi-0} \tag{95}$$

(see Fig 12b).

5.5 Extended problem 1 and locality

As well as in the case of a string, the finiteness of the wave propagation speed leads to the *local dependence* of the solution of the problem (73)–(75) on the coefficients A, B. Namely, by the arguments quite analogous to ones used in sec 2.5, the following extension of the problem 1 turns out to be a well-posed problem:

$$\rho u_{tt} - u_{xx} + Au_x + Bu = 0, \qquad (x,t) \in \Delta^{2T} \tag{96}$$

$$u\,|_{t<\frac{x}{c_1}} = 0 \tag{97}$$

$$u|_{x=0} = f, \qquad 0 \leq t \leq 2T\,, \tag{98}$$

where $\Delta^{2T} := \left\{(x,t) \mid 0 < x < c_1 T,\ 0 < t < 2T - \frac{x}{c_1}\right\}$ (see the shaded domain on Fig 13). The solution u^f of the problem (96)–(98) is determined

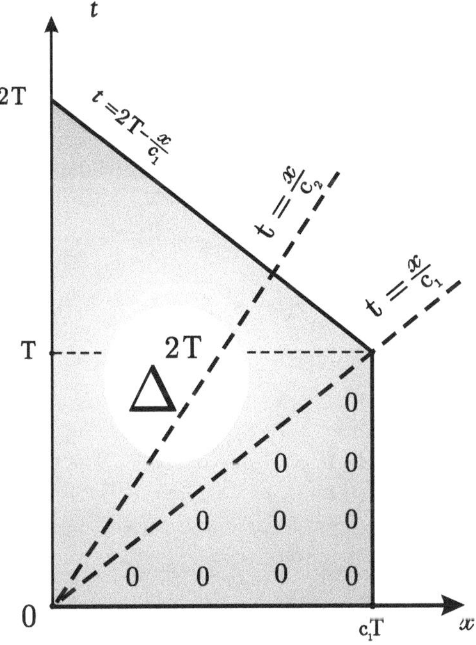

Figure 13. Extended problem

by the values of the coefficients $A, B\,|_{0<x<c_1T}$ only (do not depend on the behavior of $A, B\,|_{x \geq c_1 T}$).

Solving the matrix problem

$$\rho U_{tt} - U_{xx} + AU_x + BU = 0, \qquad (x,t) \in \Delta^{2T}$$

$$U\,|_{t<\frac{x}{c_1}} = 0$$

$$U\,|_{x=0} = \delta(t)\begin{pmatrix} 1 & 0 \\ 0 & 1 \end{pmatrix}, \qquad 0 \leq t \leq 2T\,,$$

we get the fundamental solution of the extended problem. As is easy to see, for times $t < T$ it coincides with the above-introduced fundamental solution of the problem 1, whereas the Duhamel representation

$$u^f(x,t) = U(x,t) * f(t) =$$
$$\begin{pmatrix} f_1(t-\frac{x}{c_1}) \\ f_2(t-\frac{x}{c_2}) \end{pmatrix} + \int_0^{t-\frac{x}{c_1}} W(x,t-s) \begin{pmatrix} f_1(s) \\ f_2(s) \end{pmatrix} ds\,, \qquad (x,t) \in \Delta^{2T} \tag{99}$$

holds and extends (91).

5.6 Matrix reply function

Choose a smooth control f provided $f(0) = 0$. Differentiating in (99) w.r.t. x and putting $x = 0$, one can get the representation

$$u^f_x(0,t) = U_x(0,t) * f(t) =$$
$$-\sqrt{\rho}\, f'(t) + \omega\, f(t) + \int_0^t r(t-s)\, f(s)\, ds\,, \qquad 0 < t < 2T\,, \tag{100}$$

where

$$\sqrt{\rho} = \begin{pmatrix} \sqrt{\rho_1} & 0 \\ 0 & \sqrt{\rho_2} \end{pmatrix}, \quad \omega := \begin{pmatrix} 0 & -\frac{c_1}{c_1+c_2}\, a(0) \\ \frac{c_2}{c_1+c_2}\, a(0) & 0 \end{pmatrix}$$

are the constant matrices, whereas

$$r(t) = \begin{pmatrix} r_{11}(t) & r_{12}(t) \\ r_{21}(t) & r_{22}(t) \end{pmatrix}, \qquad 0 < t < 2T$$

is a smooth matrix-function [13], which is said to be a *reply function.* It will play a central role in the inverse problem for the beam. By the locality principle, the reply function $r\,|_{0<t<2T}$ is determined by the parts $A, B\,|_{0<x<c_1 T}$ of the coefficients.

We omit the proof of (100) and recommend the reader to fill this gap himself. The helpful remark is that to derive (100) one needs to represent

$$u^f_x(0,t) = \langle \text{see (91)} \rangle = U_x(0,t) * f(t)\,, \qquad 0 < t < 2T\,,$$

[13] The case $r(+0) \neq 0$ is possible, i.e., the $r(t)$ may have a jump at $t = 0$ only.

find $U_x(x,t)$ by differentiating the representation (90) in the sense of distributions, put there $x=0$, and substitute to this convolution. The calculations are rather cumbersome and must be implemented accurately [14].

In addition, the matrix reply function possesses the following important property. As can be shown, the self-adjointness condition (76) implies

$$r^{\sharp}(t) = r(t)\,, \qquad t \geq 0\,. \tag{101}$$

To prove this relation one can just apply the trick used in [1], Lemma 2.2, p 433.

5.7 Slow waves

Here we describe a nice physical effect, which is a specific feature of two-velocity systems. This effect will play a key role in solving the inverse problem.

Return to the system (73)–(75). If $A=0$ and B is diagonal (i.e., $b_{12} = b_{21} = 0$), then the problem is decoupled to a pair of the scalar problems for the components $u_1^f(x,t)$ and $u_2^f(x,t)$. The modes $\begin{pmatrix} u_1^f \\ 0 \end{pmatrix}$ and $\begin{pmatrix} 0 \\ u_2^f \end{pmatrix}$ do not interact and propagate in the channels independently with the velocities c_1 and c_2 respectively, the second (slow) mode vanishing for $t < \frac{x}{c_2}$.

An interesting and rather unexpected fact is that the *slow waves* satisfying

$$u^f \mid_{t<\frac{x}{c_2}} = 0 \tag{102}$$

do ever exist: in spite of the inter-channel connection, a certain mixture of the modes can propagate along the beam with the slow velocity c_2! Consider the problem (73)–(75) with $T=\infty$; as is shown in [10], [12], there exists a unique smooth function $l(t)$, $t \geq 0$ such that the solution u^f satisfies (102) if and only if the components of the control $f = \begin{pmatrix} f_1 \\ f_2 \end{pmatrix}$ are linked as

$$f_1(t) = (l * f_2)\,(t) = \int_0^t l(t-s)\, f_2(s)\, ds\,, \qquad t \geq 0\,. \tag{103}$$

In particular, taking $f_2 = \delta(t)$ we get $f_1 = l(t)$, whereas the corresponding wave $u^{\begin{pmatrix} l \\ \delta \end{pmatrix}}$ is supported in the shaded space-time domain on Fig 13a. It is

[14] In the paper [12], the term $\omega f(t)$ is missed by mistake!

the wave, which is a mixture (of the fast and slow modes) propagating with the velocity c_2.

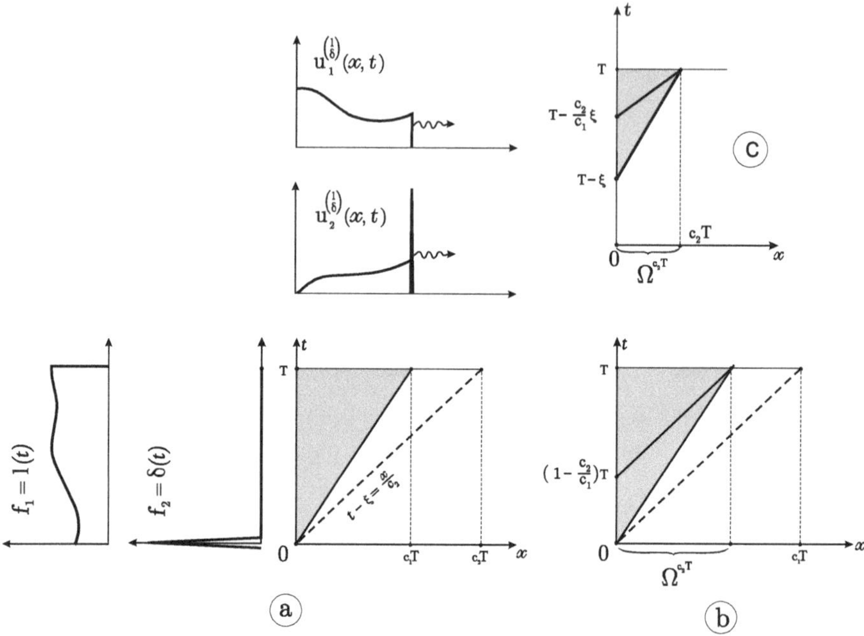

Figure 14. Slow waves

We say the function l that governs this effect to be a *delaying function.* It depends on the coefficients locally: for any $T > 0$, the values $l\mid_{0<t<(1-\frac{c_2}{c_1})T}$ are determined by the coefficients $A, B\mid_{0<x<c_2T}$ (see Fig 14b). Therefore, in the problem (73)–(75) with a finite $T > 0$, to produce the wave satisfying (102) one needs to impose the connection (103) between the control components on the interval $0 < t < (1 - \frac{c_2}{c_1})T$ only, whereas $f_1, f_2\mid_{(1-\frac{c_2}{c_1})T<t<T}$ can be arbitrary: varying these values, one cannot violate (102) just because the maximal wave propagation speed is c_1.

Fix a positive $\xi \leq T$. Combining the above mentioned properties of slow waves with the delay relation (93), we easily conclude that the solution u^f satisfies

$$u^f\mid_{t<T-\xi+\frac{x}{c_2}} = 0 \tag{104}$$

if and only if the control f satisfies

$$f\mid_{t<T-\xi}=0 \qquad \text{and}$$

$$f_1(t)=\int\limits_{T-\xi}^{t} l(t-s)\,f_2(s)\,ds\,, \quad T-\xi<t<T-\frac{c_2}{c_1}\xi\,. \tag{105}$$

Such solutions are supported in the domain shaded of Fig 14c and can be regarded as "delayed slow waves".

6 Beam as dynamical system

6.1 System α^T

Here the problem (73)–(75) is regarded in terms of the control theory. We endow it with standard attributes of a dynamical system: spaces and operators. The system is denoted by α^T.

Outer space By angle brackets we denote the standard inner product in $\mathbb{R}^2$: for $a=\begin{pmatrix}a_1\\a_2\end{pmatrix}$, $b=\begin{pmatrix}b_1\\b_2\end{pmatrix}$ we put $\langle a,b\rangle:=a_1b_1+a_2b_2$.

The Hilbert space of controls (inputs) $\mathcal{F}^T:=L_2\left((0,T);\mathbb{R}^2\right)$ with the inner product

$$(f,g)_{\mathcal{F}^T}=\int\limits_0^T\langle f(t),g(t)\rangle\,dt$$

is said to be an *outer space* of the system α^T. It contains an increasing family of subspaces

$$\mathcal{F}^{T,\xi}:=\left\{f\in\mathcal{F}^T\mid\ f|_{t<T-\xi}=0\right\}\,,\qquad 0\le\xi\le T$$

consisting of the delayed controls ($T-\xi$ is the value of the delay, ξ is an action time). Note that $\mathcal{F}^{T,0}=\{0\}$ and $\mathcal{F}^{T,T}=\mathcal{F}^T$.

A specific feature of a two-velocity system is that there is one more increasing family of subspaces in $\mathcal{F}^T$ related to slow waves:

$$\mathcal{F}_l^{T,\xi}:=$$

$$\left\{f=\begin{pmatrix}f_1\\f_2\end{pmatrix}\in\mathcal{F}^{T,\xi}\ \middle|\ f_1(t)=\int\limits_{T-\xi}^{t}l(t-s)\,f_2(s)\,ds\,,\quad T-\xi<t<T-\frac{c_2}{c_1}\xi\right\}. \tag{106}$$

In accordance with (104) and (105), the wave u^f satisfies (104) if and only if the control f belongs to the class $\mathcal{F}_l^{T,\xi}$. Also, notice that $\mathcal{F}_l^{T,\xi} \subset \mathcal{F}^{T,\xi}$.

Inner space Recall that by Ω^a we denote the interval $(0,a)$ of the semi-axis $x>0$. The space $\mathcal{H}^{c_1T} := L_{2,\rho}(\Omega^{c_1T})$ with the inner product

$$(u,v)_{\mathcal{H}^{c_1T}} = \int\limits_{\Omega^{c_1T}} \langle \rho u(x), v(x)\rangle\, dx = \int\limits_0^{c_1T} [\rho_1 u_1(x)v_1(x) + \rho_2 u_2(x)v_2(x)]\, dx$$

is called an *inner space* of the system α^T. For each $f \in \mathcal{F}^T$, at any moment t the wave $u^f(\,\cdot\,,t)$ is supported in Ω^{c_1T} (see (92)) and, hence, can be regarded as a time-dependent element of $\mathcal{H}^{c_1T}$. In control theory, $u^f(\,\cdot\,,t)$ is referred to as a *state* of the system at the moment t. So, $\mathcal{H}^{c_1T}$ is a space of states.

The inner space contains two increasing families of subspaces

$$\mathcal{H}^{c_i\xi} := \left\{y \in \mathcal{H}^{c_1T} \mid \; y|_{x>c_i\xi} = 0\right\} \qquad 0 \le \xi \le T, \; i=1,2$$

of vector-functions supported in the intervals $\Omega^{c_1\xi}$ and $\Omega^{c_2\xi}$ respectively. Notice that $\mathcal{H}^{c_2\xi} \subset \mathcal{H}^{c_1\xi}$ for every $\xi>0$. In accordance with (92), $\Omega^{c_1\xi}$ can be regarded as an interval of the beam filled with waves at the moment $t=\xi$, whereas by (102) $\Omega^{c_2\xi}$ is the interval filled with slow waves.

Control operator The input $\mapsto$ state correspondence of the system α^T is realized by a *control operator* $W^T : \mathcal{F}^T \to \mathcal{H}^{c_1T}$,

$$W^T f := u^f(\,\cdot\,,T)$$

which creates the waves. By (91), it can be represented in the form

$$\left(W^T f\right)(x) = \begin{pmatrix} f_1(T-\frac{x}{c_1}) \\ f_2(T-\frac{x}{c_2}) \end{pmatrix} + \int\limits_0^{T-\frac{x}{c_1}} W(x,T-s)\, f(s)\, ds\,, \tag{107}$$

which shows that W^T is a bounded operator.

Combining (92) with (93), it is easy to see that the condition $f\,|_{t<T-\xi}=0$ implies $u^f(\,\cdot\,,T)\,|_{x>c_1\xi}=0$, i.e., the embedding

$$W^T\mathcal{F}^{T,\xi} \subset \mathcal{H}^{c_1\xi}\,, \qquad 0 \le \xi \le T \tag{108}$$

holds. In the mean time, taking $f \in \mathcal{F}_l^{T,\xi}$ and putting $t=T$ in (104) we see that $u^f(\,\cdot\,,T)\,|_{x>c_2\xi}=0$, i.e., $u^f(\,\cdot\,,T)$ is supported in $\Omega^{c_2\xi}$ and hence we have the embedding

$$W^T\mathcal{F}_l^{T,\xi} \subset \mathcal{H}^{c_2\xi}\,, \qquad 0 \le \xi \le T \tag{109}$$

We shall see that the character of these relations is different: the first embedding is strict, whereas the second one is in fact an equality.

Response operator An input $\mapsto$ output correspondence in the system α^T is realized by a *response operator* $R^T : \mathcal{F}^T \to \mathcal{F}^T$ which is defined on controls f of the Sobolev class $H^1\left([0,T];\mathbb{R}^2\right)$ [15] provided $f(0)=0$ and acts by the rule

$$\left(R^T f\right)(t) := u^f_x(0,t) = \begin{pmatrix} \frac{\partial u^f_1}{\partial x}(0,t) \\ \frac{\partial u^f_2}{\partial x}(0,t) \end{pmatrix}, \quad 0<t<T\,.$$

Using (100) for times $0<t<T$, we can represent

$$\left(R^T f\right)(t) = -\sqrt{\rho}\, f'(t)+\omega\, f(t)+\int\limits_0^t r(t-s)\, f(s)\, ds\,, \qquad 0<t<T\,. \quad (110)$$

It is easy to check that the adjoint operator $\left(R^T\right)^*$ is well defined on controls $f\in H^1\left([0,T];\mathbb{R}^2\right)$ provided $f(T)=0$ and acts in the outer space by the rule

$$\left(\left(R^T\right)^* f\right)(t) =$$

$$\sqrt{\rho}\, f'(t)+\omega^\sharp\, f(t)+\int\limits_t^T r^\sharp(s-t)\, f(s)\, ds\,, \qquad 0<t<T\,. \quad (111)$$

With the extended problem 1 (see (96)–(98)) one associates the extended response operator R^{2T}, which acts in the space $\mathcal{F}^{2T} := L_2\left((0,2T);\mathbb{R}^2\right)$ on controls f of the class $H^1\left([0,2T];\mathbb{R}^2\right)$ provided $f(0)=0$ by the rule

$$\left(R^{2T} f\right)(t) := u^f_x(0,t), \quad 0<t<2T\,,$$

where u^f is the solution of (96)–(98). In accordance with (100), the representation

$$\left(R^{2T} f\right)(t) = -\sqrt{\rho}\, f'(t)+\omega\, f(t)+\int\limits_0^t r(t-s)\, f(s)\, ds\,, \qquad 0<t<2T \quad (112)$$

is valid. By the locality principle, along with the reply matrix-function r the operator R^{2T} is determined by the part $A, B\,|_{\Omega^{c_1 T}}$ of the coefficients.

[15] For controls of this class, both components belong to the scalar Sobolev space $H^1[0,T]$.

Connecting Operator Since the control operator W^T acts from a Hilbert space $\mathcal{F}^T$ to a Hilbert space $\mathcal{H}^{c_1T}$, its adjoint $\left(W^T\right)^*$ acts from $\mathcal{H}^{c_1T}$ to $\mathcal{F}^T$. By this, an operator

$$C^T := \left(W^T\right)^* W^T$$

is well defined. It acts in the outer space $\mathcal{F}^T$ and is said to be a *connecting operator* of the system α^T. Since W^T is a bounded operator, C^T is also bounded. By the definition, one has

$$\begin{aligned} \left(C^T f, g\right)_{\mathcal{F}^T} &= \left(\left(W^T\right)^* W^T f, g\right)_{\mathcal{F}^T} = \left(W^T f, W^T g\right)_{\mathcal{H}^{c_1T}} \\ &= \left(u^f(\cdot,T), u^g(\cdot,T)\right)_{\mathcal{H}^{c_1T}}, \end{aligned} \tag{113}$$

i.e., C^T connects the metrics of the outer and inner spaces.

The following remarkable fact occurs: *if the self-adjointness condition (76) is fulfilled* then the connecting operator can be explicitly and simply represented via the response operator, the representation being of the same form as in the scalar case:

$$C^T = -\frac{1}{2}\left(S^T\right)^* R^{2T} J^{2T} S^T, \tag{114}$$

where the operators S^T and J^{2T} are of the same meaning as in the case of the string: see (37). Moreover, the representation through the reply function

$$\left(C^T f\right)(t) = \sqrt{\rho}\, f(t) + \int_0^T c^T(t,s)\, f(s)\, ds\,, \qquad 0 < t < T\,, \tag{115}$$

with the matrix kernel

$$c^T(t,s) := \frac{1}{2}\int_{|t-s|}^{2T-t-s} r(\eta)\, d\eta \qquad 0 < s,t < T$$

also remains in force [16]. To establish (114) one can just repeat the derivation of (37) described in Appendix II of Part I, i.e., to introduce the Blagovestchenskii function

$$b(s,t) := \int_0^\infty \langle \rho u^{f_-}(x,s), u^g(x,t)\rangle\, dx, \qquad (s,t) \in B^{2T}$$

[16] Recall that in the self-adjoint case the reply function is symmetric: see (101).

and then derive the equation $b_{tt} - b_{ss} = F$ by integration by parts with the use of the equalities

$$\int_0^\infty \left\langle \left[\frac{d^2}{dx^2} - A(x)\frac{d}{dx} - B(x)\right] u(x), v(x) \right\rangle dx =$$

$$\left[\langle u_x, v\rangle - \langle u, v_x\rangle - \langle u, A^\sharp v\rangle\right]\Big|_{x=0}^{\infty} +$$

$$\int_0^\infty \left\langle u(x), \left[\frac{d^2}{dx^2} + A^\sharp(x)\frac{d}{dx} + \left(A^\sharp_x(x) - B^\sharp(x)\right)\right] v(x) \right\rangle dx =$$

$$- \langle u_x(0), v(0)\rangle + \langle u(0), v_x(0)\rangle + \langle u(0), A^\sharp(0)v(0)\rangle +$$

$$\int_0^\infty \left\langle u(x), \left[\frac{d^2}{dx^2} - A(x)\frac{d}{dx} - B(x)\right] v(x) \right\rangle dx \qquad (116)$$

for smooth vector-functions u, v vanishing for large x's. The reader is strongly recommended to restore all details of the derivation.

The significance of the formula (114) for the inverse problem is the same as in the scalar case: we shall see that it is the relation, which enables the external observer to visualize the waves in the beam through the measurements at the endpoint $x = 0$.

The diagram on Fig 6 is also quite available for the system α^T corresponding to a beam.

6.2 Controllability

The relevant statement of the boundary control problem (BCP) for a beam is the same as for a string: *given a vector-function* $y \in \mathcal{H}^{c_1 T}$ *to find a control* $f \in \mathcal{F}^T$ such that

$$u^f(x, T) = y(x), \qquad x \in \Omega^{c_1 T}. \qquad (117)$$

However we shall see that the character of controllability of a beam differs essentially from the one of a string.

By analogy to the scalar case, introduce a *reachable set*

$$\mathcal{U}^T := \{u^f(\cdot, T) \mid f \in \mathcal{F}^T\} = W^T \mathcal{F}^T \subset \mathcal{H}^{c_1 T}.$$

The principal fact is that the latter embedding is now strict: in contrast to the property (43), on the beam we have

$$\mathcal{U}^T \neq \mathcal{H}^{c_1 T}, \qquad T > 0. \qquad (118)$$

Moreover, the subspace $\mathcal{H}^{c_1T} \ominus \mathcal{U}^T \neq \{0\}$ turns out to be of *infinite dimension*, i.e., in the inner space $\mathcal{H}^{c_1T}$ there exist infinitely many vector-functions y orthogonal to all waves[17]. So, on a beam the local boundary controllability does not occur!

Such a fact should be expected. Indeed if $A = 0$ and B is diagonal, the channels do not interact and, for any control f, the second component $u_2^f(\cdot, T)$ is supported in the interval Ω^{c_2T} filled with slow modes (see Fig 16a)

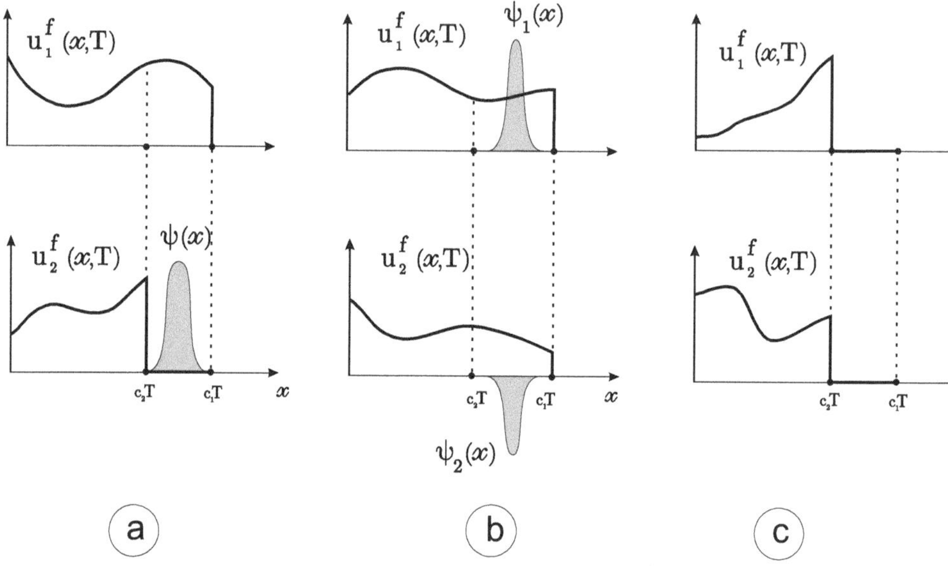

Figure 15. Lack of controllability in Ω^{c_1T}

Therefore, the vector-function $y = \begin{pmatrix} 0 \\ \psi \end{pmatrix}$, where ψ is an arbitrary scalar function supported in $\Omega^{c_1T} \setminus \Omega^{c_2T}$, is automatically orthogonal to the waves (see Fig 16a) and hence do not belong to the reachable set.

The same effect occurs if the channels interact, what can be explained as follows. For any wave $u^f(\cdot, T)$, the precursor $u_2^f(\cdot, T)\,|_{c_2T<x<c_1T}$ in the slow channel is induced by the head part of the component $u_1^f(\cdot, T)\,|_{c_2T<x<c_1T}$ in the fast channel and *is determined by this head part.* More precisely, there exists an operator $K^T : L_2(c_2T, c_1T) \to L_2(c_2T, c_1T)$ that maps the head part to the precursor, the operator being one and the same for all controls:

[17]Such functions are referred to as *unreachable states.*

see [11] for detail. By this, for the functions ψ_1, ψ_2 supported in $\Omega^{c_1T} \setminus \Omega^{c_2T}$ we have

$$\left(u^f(\,\cdot\,,T), \begin{pmatrix}\psi_1\\ \psi_2\end{pmatrix}\right)_{\mathcal{H}^{c_1T}} = \int_{c_2T}^{c_1T} \left\langle \rho \begin{pmatrix} u_1^f(x,T)\\ u_2^f(x,T)\end{pmatrix}, \begin{pmatrix}\psi_1(x)\\ \psi_2(x)\end{pmatrix}\right\rangle dx =$$

$$\int_{c_2T}^{c_1T} \left\langle \rho \begin{pmatrix} u_1^f(x,T)\\ \left(K^T u_1^f(\,\cdot\,,T)\right)(x)\end{pmatrix}, \begin{pmatrix}\psi_1(x)\\ \psi_2(x)\end{pmatrix}\right\rangle dx =$$

$$\int_{c_2T}^{c_1T} \left[\rho_1\, u_1^f(x,T)\, \psi_1(x) + \rho_2 \left(K^T u_1^f(\,\cdot\,,T)\right)(x)\, \psi_2(x)\right] dx =$$

$$\int_{c_2T}^{c_1T} \left[\rho_1\, \psi_1(x) + \rho_2 \left(\left(K^T\right)^* \psi_2\right)(x)\right] u_1^f(x,T)\, dx\,.$$

Therefore, taking $\psi_1 = -\frac{\rho_2}{\rho_1}\left(K^T\right)^* \psi_2$, we get

$$\left(u^f(\,\cdot\,,T), \begin{pmatrix}\psi_1\\ \psi_2\end{pmatrix}\right)_{\mathcal{H}^{c_1T}} = 0$$

for all $f \in \mathcal{F}^T$, i.e., $\begin{pmatrix}\psi_1\\ \psi_2\end{pmatrix} \perp \mathcal{U}^T$ (see Fig 16b).

A remarkable fact is that the local controllability can be restored if one uses the slow waves. Namely, introduce the reachable set of the form

$\mathcal{U}_l^T := W^T \mathcal{F}_l^{T,T} = \langle$ see (106) for $\xi = T \rangle =$

$$\left\{ u^f(\,\cdot\,,T) \;\middle|\; f = \begin{pmatrix} f_1\\ f_2\end{pmatrix}:\; f_1(t) = \int_0^t l(t-s)\, f_2(s)\, ds\,, \quad 0 < t < T - \frac{c_2}{c_1}T \right\}.$$

By the definition and properties of the slow waves, each $u^f(\,\cdot\,,T)$ is supported in Ω^{c_2T} (see Fig 14b and Fig 15c) and hence we have an embedding $\mathcal{U}_l^T \subset \mathcal{H}^{c_2T}$. However this embedding is in fact the equality:

$$\mathcal{U}_l^T = \mathcal{H}^{c_2T}\,, \qquad T > 0\,. \tag{119}$$

To prove (119) let us consider an auxiliary problem

$$\rho v_{tt} - v_{xx} + A v_x + B v = 0, \qquad 0 < x < c_2T,\; 0 < t < T \tag{120}$$

$$v\,|_{t<\frac{x}{c_2}} = 0, \tag{121}$$

$$v\,|_{t=T} = y\,, \qquad 0 \le x \le c_2T \tag{122}$$

with $y \in \mathcal{H}^{c_2T}$. Such a problem is well posed. Indeed, it differs from the problem (73)–(75) just by the role of variables x and t: in (120)–(122) it is natural to regard x as a time, whereas t plays the role of a space coordinate (see Fig 16). Let $v = v^y(x,t)$ be the solution; denote $f :=$

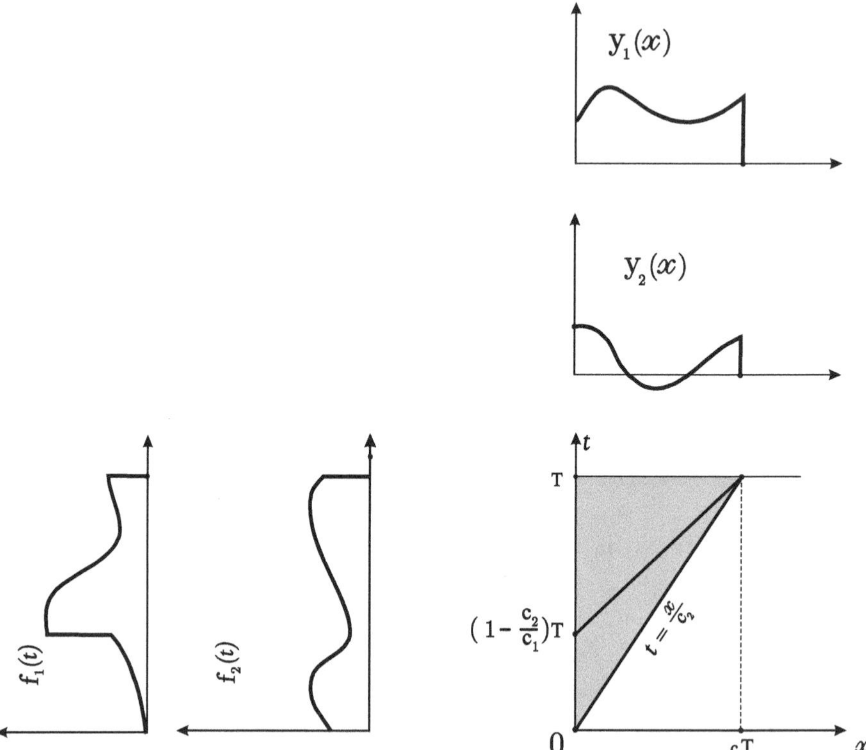

Figure 16. Auxiliary problem. Controllability in Ω^{c_2T}

$v^y(0,t)$, $0<t<T$. Comparing (120)–(122) with (73)–(75), we easily see that $u^f(x,t) = v^y(x,t)$ for all $x \in \Omega^{c_2T}$, $0<t<T$ and the relation

$$u^f(x,T) = y(x), \qquad x \in \Omega^{c_2T} \tag{123}$$

holds. Also, since the solution u^f vanishes as $t < \frac{x}{c_2}$ (see (121)), it is a *slow wave* and hence the control f, which this wave is produced by, has to belong to the subspace $\mathcal{F}_l^{T,T}$.

Resuming the aforesaid, we conclude that for any vector-function $y \in \mathcal{H}^{c_2T}$ one can find a unique control $f \in \mathcal{F}_l^{T,T}$ such that (123) is fulfilled.

Hence, for such y's, the BCP (117) is uniquely solvable and we arrive at (119). Thus, by making use of controls of the class $\mathcal{F}_l^{T,T}$ we restore the local boundary controllability of the system α^T *but on the "slow" intervals* $\Omega^{c_2 T}$ *only!*

For intermediate times $t = \xi$, let us introduce the "slow" reachable sets $\mathcal{U}_l^\xi := W^T \mathcal{F}_l^{T,\,\xi}$. With regard to the delay relation (93), they can be also specified as $\mathcal{U}_l^\xi = \{u^f(\,\cdot\,,\xi) \mid f \in \mathcal{F}_l^{T,T}\}$. Since the final moment T in the equality (119) is arbitrary, it can be written in the form

$$W^T \mathcal{F}_l^{T,\,\xi} = \mathcal{H}^{c_2 \xi}\,, \qquad 0 \le \xi \le T\,, \tag{124}$$

which is convenient for what follows. In operator terms, (124) means that the control operator maps every subspace $\mathcal{F}_l^{T,\,\xi}$ onto the subspace $\mathcal{H}^{c_2\xi}$ isomorphically, whereas the BCP (117) for $y \in \mathcal{H}^{c_2\xi}$ is solved by $f = \left(W^T\right)^{-1} y \in \mathcal{F}_l^{T,\,\xi}$.

6.3 Wave basis

Here, taking into account the specific character of the controllability of a beam, we construct the *slow wave bases* in the reachable subspaces $\mathcal{H}^{c_2\xi}$ and the corresponding projection operators.

1. Fix a positive $\xi \le T$ and choose a basis of controls $\{f_j^\xi\}_{j=1}^\infty$ in the subspace $\mathcal{F}_l^{T,\,\xi}$. By (124), the corresponding waves $\{u^{f_j^\xi}(\,\cdot\,,T)\}_{j=1}^\infty$ constitute a basis of the subspace $\mathcal{H}^{c_2\xi}$.
2. Apply to the basis $\{f_j^\xi\}_{j=1}^\infty$ the Schmidt orthogonalization process w.r.t. the bilinear form $\left(C^T f, g\right)_{\mathcal{F}^T}$:

$$\tilde g_1^\xi := f_1^\xi\,, \qquad g_1^\xi := \frac{\tilde g_1^\xi}{\sqrt{\left(C^T \tilde g_1^\xi\,, \tilde g_1^\xi\right)_{\mathcal{F}^T}}}\,;$$

$$\tilde g_2^\xi := f_2^\xi - \left(C^T f_2^\xi\,, g_1^\xi\right)_{\mathcal{F}^T} g_1^\xi\,, \qquad g_2^\xi := \frac{\tilde g_1^\xi}{\sqrt{\left(C^T \tilde g_2^\xi\,, \tilde g_2^\xi\right)_{\mathcal{F}^T}}}\,;$$

$$\cdots\cdots\cdots\cdots\cdots\cdots\cdots\cdots\cdots\cdots\cdots\cdots$$

$$\tilde g_k^\xi := f_k^\xi - \sum_{j=1}^{k-1} \left(C^T f_k^\xi\,, g_j^\xi\right)_{\mathcal{F}^T} g_j^\xi, \qquad g_k^\xi := \frac{\tilde g_k^\xi}{\sqrt{\left(C^T \tilde g_k^\xi\,, \tilde g_k^\xi\right)_{\mathcal{F}^T}}}\,;$$

$$\cdots\cdots\cdots\cdots\cdots\cdots\cdots\cdots\cdots\cdots\cdots\cdots\,. \tag{125}$$

As a result, we get a new system of controls $\{g_k^\xi\}_{k=1}^\infty$, which is also a basis in $\mathcal{F}_l^{T,\,\xi}$ and satisfies

$$\left(C^T g_k^\xi\,,\,g_l^\xi\right)_{\mathcal{F}^T} \;=\; \delta_{kl} \qquad k,l=1,2,\dots \tag{126}$$

by construction.

The corresponding waves $\{u_k^\xi\}_{k=1}^\infty\,,\ \ u_k^\xi := u^{g_k^\xi}(\,\cdot\,,T) = W^T g_k^\xi$ form an orthonormalized basis in the subspace $\mathcal{H}^{c_2\xi}$. Indeed, we have

$$\left(u_k^\xi\,,\,u_l^\xi\right)_{\mathcal{H}^{cT}} \;=\; \langle\text{see (113)}\rangle \;=\; \left(C^T g_k^\xi\,,\,g_l^\xi\right)_{\mathcal{F}^T} \;=\; \langle\text{see (126)}\rangle \;=\; \delta_{kl} \qquad k,l=1,2,\dots\,.$$

We say $\{u_k^\xi\}_{k=1}^\infty$ to be a *slow wave basis* in $\mathcal{H}^{c_2\xi}$.

Of course, from the computational point of view, this orthogonalization process is more complicated than its scalar analog (see sec 3.3) because one deals with the vector-functions. However, as well as for a string, to construct the system $\{g_k^\xi\}_{k=1}^N$ is in fact to invert the Gram matrix $\left\{(C^T f_i^\xi, f_j^\xi)_{\mathcal{F}^T}\right\}_{i,j=1}^N$ for large N's. Since C^T is a boundedly invertible operator, the inversion turns out to be a well-posed procedure[18].

3. The basis $\{g_k^\xi\}_{k=1}^\infty \subset \mathcal{F}_l^{T,\,\xi}$ determines an operator $\mathcal{P}_l^\xi$, which acts in the external space $\mathcal{F}^T$ by the rule

$$\mathcal{P}_l^\xi f \;=\; \sum_{k=1}^{\infty}\left(C^T f\,,\,g_k^\xi\right)_{\mathcal{F}^T}\,g_k^\xi \tag{127}$$

and will work in the inverse problem. Note that, for any control f, its image $\mathcal{P}_l^\xi f$ belongs to the subspace $\mathcal{F}_l^{T,\,\xi}$. Moreover, it can be shown that $\mathcal{P}_l^\xi$ is a skew projection in $\mathcal{F}^T$ onto $\mathcal{F}_l^{T,\,\xi}$ [19].

[18] Such a well-posedness is a specific feature of the one-dimensional (w.r.t. the space variable x) problems: in the multidimensional case the situation is much worse and complicated (see, e.g., [7]).

[19] In the mean time, it can be interpreted as the orthogonal projection w.r.t. the inner product $(C^T f, g)_{\mathcal{F}^T}$: see [12].

6.4 Truncation and AF

As in the previous section, we keep $\xi \in (0, T]$ fixed. In the inner space $\mathcal{H}^{c_1 T}$, introduce the operation $P^{c_2\xi}$

$$\left(P^{c_2\xi} y\right)(x) := \begin{cases} y(x)\,, & x \in \Omega^{c_2\xi} \\ 0\,, & x \in \Omega^{c_1 T} \setminus \Omega^{c_2\xi} \end{cases}$$

that truncates functions onto the subinterval $\Omega^{c_2\xi} \subset \Omega^{c_1 T}$ filled with slow waves. It is easy to recognize that $P^{c_2\xi}$ is the orthogonal projection in $\mathcal{H}^{c_1 T}$ onto the subspace $\mathcal{H}^{c_2\xi}$. Therefore, expanding over the slow wave basis, we can represent the truncated function in the form of the Fourier series:

$$P^{c_2\xi}\, y \;=\; \sum_{k=1}^{\infty} \left(y\,,\, u_k^\xi\right)_{\mathcal{H}^{c_1 T}} u_k^\xi\,. \tag{128}$$

The truncation projector is related to the above-introduced operator (127) as follows: for any $f \in \mathcal{F}^T$ the equality

$$P^{c_2\xi} W^T f \;=\; W^T \mathcal{P}_l^\xi f \tag{129}$$

is valid. Indeed, taking $y = W^T f$ in (128) and recalling that $u_k^\xi := W^T g_k^\xi$, we have

$$\begin{aligned} P^{c_2\xi} W^T f \;&=\; \sum_{k=1}^{\infty} \left(W^T f\,,\, W^T g_k^\xi\right)_{\mathcal{H}^{c_1 T}} W^T g_k^\xi \;=\; \langle \text{see (113)} \rangle \\ &= W^T \sum_{k=1}^{\infty} \left(C^T f\,,\, g_k^\xi\right)_{\mathcal{F}^T} g_k^\xi \;=\; \langle \text{see (127)} \rangle = W^T \mathcal{P}_l^\xi f\,. \end{aligned}$$

Choose a control $f \in \mathcal{F}^T$ to be a continuous vector-function and write (129) in the form

$$P^{c_2\xi} u^f(\,\cdot\,, T) \;=\; u^{\mathcal{P}_l^\xi f}(\,\cdot\,, T)\,. \tag{130}$$

The truncated wave in the l.h.s. is a piece-wise continuous function, which vanishes in $\Omega^{c_1 T} \setminus \Omega^{c_2\xi}$ and has a jump at the point $x = c_2\xi$, the value of the jump being equal to $u^f(c_2\xi, T)$. On the other hand, by (130) [20] the truncated wave is also *a wave* produced by the control $\mathcal{P}_l^\xi f \in \mathcal{F}_l^{T,\,\xi}$, which vanishes for $t < T - \xi$ and has the certain jumps at the moments

[20] and, eventually, by the controllability (124)!

$t=T-\xi$ and $t=T-\frac{c_2}{c_1}\xi$. The control and wave jumps are related through (95). Applying this relation for the control $\mathcal{P}_l^\xi f$, we easily arrive at the key equalities

$$u_1^f(c_2\xi,T)=\left[\left(\mathcal{P}_l^\xi f\right)_1(t)\right]\Big|_{t=T-\frac{c_2}{c_1}\xi-0}^{t=T-\frac{c_2}{c_1}\xi+0},$$

$$u_2^f(c_2\xi,T)=\left[\left(\mathcal{P}_l^\xi f\right)_2(t)\right]\Big|_{t=T-\xi+0}, \qquad 0<\xi\le T, \qquad (131)$$

which represent the values of the wave $u^f(\,\cdot\,,T)$ on the interval Ω^{c_2T} filled with slow waves, through the operator $\mathcal{P}_l^\xi$.

As in the scalar case, we call the representation (131) the *amplitude formula* (AF). It represents a wave through the amplitudes of jumps of control, the jumps appearing as result of projecting the control on the subspaces $\mathcal{F}^{T,\xi}$ by the operators $\mathcal{P}^\xi$. The background of the AF is geometrical optics; various versions of this formula play the key role in the BC-method, including its multidimensional variants [7], [12], [6].

6.5 Special BCP

In this section we deal with the problem (117) with a special r.h.s. y. The self-adjointness conditions (76) are assumed fulfilled.

Consider a Cauchy problem

$$-p''+Ap'+Bp=0, \qquad x>0 \qquad (132)$$

$$p\,|_{x=0}=\alpha, \quad p'\,|_{x=0}=\beta \qquad (133)$$

with the given vectors $\alpha,\beta\in\mathbb{R}^2$. By standard ODE theory, such a problem has a unique smooth solution (vector-function) $p=p_{\alpha\beta}(x)$. In what follows we assume α,β to be fixed and write just $p(x)$.

The analog of the scalar special BCP (55) for the beam has to take into account the character of its controllability, i.e., the relation (119). Since the waves produced by controls $f\in\mathcal{F}_l^{T,T}$ are slow, at the final moment $t=T$ they are supported in Ω^{c_2T}. By this, the relevant set up is as follows: *to find a control $f\in\mathcal{F}_l^{T,T}$ satisfying*

$$u^f(x,T)=p(x), \qquad x\in\Omega^{c_2T}. \qquad (134)$$

By (124) for $\xi=T$, such a problem has a unique solution that we denote by f^T.

In the case of a beam, the representation of the function $p = \begin{pmatrix} p_1(x) \\ p_2(x) \end{pmatrix}$, which generalizes (60) and is available for the inverse problem, is the following:

$$
\begin{aligned}
&p_1(c_2\xi, T) = \\
&\left\{ \sum_{k=1}^{\infty} \left(\left[-\left(R^T\right)^* + A^\sharp(0) \right] \varkappa^T \alpha + \varkappa^T \beta \, , g_k^\xi \right)_{\mathcal{F}^T} g_k^\xi(t) \right\}_1 \Bigg|_{t=T-\frac{c_2}{c_1}\xi-0}^{t=T-\frac{c_2}{c_1}\xi+0}, \\
&p_2(c_2\xi, T) = \\
&\left\{ \sum_{k=1}^{\infty} \left(\left[-\left(R^T\right)^* + A^\sharp(0) \right] \varkappa^T \alpha + \varkappa^T \beta \, , g_k^\xi \right)_{\mathcal{F}^T} g_k^\xi(t) \right\}_2 \Bigg|_{t=T-\xi+0}, \\
&0 < \xi \le T .
\end{aligned}
\tag{135}
$$

Let us derive it.

Applying the AF (131) for $f = f^T$ we get

$$
\begin{aligned}
p_1(c_2\xi, T) &= \left[\left(\mathcal{P}_l^\xi f^T \right)_1 (t) \right] \Bigg|_{t=T-\frac{c_2}{c_1}\xi-0}^{t=T-\frac{c_2}{c_1}\xi+0}, \\
p_2(c_2\xi, T) &= \left[\left(\mathcal{P}_l^\xi f^T \right)_2 (t) \right] \Bigg|_{t=T-\xi+0}, \qquad 0 < \xi \le T ,
\end{aligned}
\tag{136}
$$

where

$$
\mathcal{P}_l^\xi f^T = \sum_{k=1}^{\infty} \left(C^T f^T , g_k^\xi \right)_{\mathcal{F}^T} g_k^\xi \tag{137}
$$

(see (127)). The coefficients of the series can be found as follows. Let us take a smooth control $g \in \mathcal{F}_l^{T,T}$ provided $g(0) = g'(0) = g''(0) = 0$, so that the corresponding wave u^g satisfies (73)–(75) in the classic sense. Such a wave is slow: it vanishes at its forward front $t = \frac{x}{c_2}$ along with derivatives; therefore, at the final moment we have

$$
u^g(x,T)\,|_{x>c_2T} = 0 , \qquad u^g(c_2T, T) = u_x^g(c_2T, T) = 0 . \tag{138}
$$

This enables one to justify the following calculations:

$$\left(C^T f^T, g\right)_{\mathcal{F}^T} = \langle \text{see (113)} \rangle = \left(u^{f^T}(\,\cdot\,,T), u^g(\,\cdot\,,T)\right)_{\mathcal{H}^{c_1 T}} =$$

$$\langle \text{see (134), (138)} \rangle = \int_0^{c_2 T} \langle \rho\, p(x)\,, u^g(x,T)\rangle\, dx =$$

$$\langle \text{by zero Cauchy data (74) for the wave } u^g \rangle =$$

$$\int_0^{c_2 T} \left\langle \rho\, p(x)\,, \int_0^T (T-t)\, u^g_{tt}(x,t)\, dt \right\rangle dx =$$

$$\int_0^T dt\,(T-t) \int_0^{c_2 T} \langle p(x)\,, \rho\, u^g_{tt}(x,t)\rangle\, dx = \langle \text{see (73)} \rangle =$$

$$\int_0^T dt\,(T-t) \int_0^{c_2 T} \langle p(x)\,,\, u^g_{xx}(x,t) - A(x) u^g_x(x,t) - B(x) u^g(x,t)\rangle\, dx =$$

$$\langle \text{see (116), (138)} \rangle =$$

$$\int_0^T (T-t)\, \left[\left\langle -p(0), \left(R^T g\right)(t)\right\rangle + \langle p'(0), g(t)\rangle + \left\langle A^\sharp(0) p(0), g(t)\right\rangle\right] dt$$

$$= \left(\left[-\left(R^T\right)^* + A^\sharp(0)\right] \varkappa^T \alpha + \varkappa^T \beta\,, g\right)_{\mathcal{F}^T}. \tag{139}$$

Thus, we have got the equality

$$\left(C^T f^T, g\right)_{\mathcal{F}^T} = \left(\left[-\left(R^T\right)^* + A^\sharp(0)\right] \varkappa^T \alpha + \varkappa^T \beta\,,\, g\right)_{\mathcal{F}^T} \tag{140}$$

for classical g's. Since such controls constitute a dense set in $\mathcal{F}^T$, the evident limit passage extends (140) to arbitrary $g \in \mathcal{F}_l^{T,T}$.

Taking $g = g_k^\xi$ in (140) and returning to (137), we obtain

$$\left(\mathcal{P}_l^\xi f^T\right)(t) =$$

$$\sum_{k=1}^{\infty} \left(\left[-\left(R^T\right)^* + A^\sharp(0)\right] \varkappa^T \alpha + \varkappa^T \beta\,, g_k^\xi\right)_{\mathcal{F}^T} g_k^\xi(t)\,, \qquad 0 < t < T\,. \tag{141}$$

Summarizing the above considerations, we substitute (141) in (136) and arrive at (135).

The significance of the formula (135) is that it expresses the function p, which is an object of the *inner space*, in intrinsic terms of the *outer space*. It is the fact, which renders this representation relevant to the inverse problem.

In the scalar case, the solution of the special BCP satisfies the Gelfand-Levitan-Krein equation (61). Does there exist the relevant analog of this equation for the beam? In other words, can one extend the Sondhi-Gopinath approach to the case of 2-velocity system? The answer is affirmative but however such an analog turns out to be of rather complicated form. Namely, the relation (140) easily implies that the control $f = f^T$ is *a unique element of the subspace* $\mathcal{F}_l^{T,T}$, *which satisfies the equation*

$$\mathcal{P}_l^T C^T f = \mathcal{P}_l^T \left\{ \left[-\left(R^T\right)^* + A^\sharp(0) \right] \varkappa^T \alpha + \varkappa^T \beta \right\} \tag{142}$$

(see [14]).

7 Inverse Problem

7.1 Statement

The external observer applies controls $f(t)\,|_{(0,2T)}$ at the endpoint $x = 0$ of the beam and measures $u_x^f(0,t)\,|_{(0,2T)}$ at the same point. As result of such measurements, the observer possesses the response operator R^{2T} (see (112))[21]. In accordance with the locality principle, the operator R^{2T} is determined by the coefficients $A, B\,|_{\Omega^{c_1 T}}$. Therefore, at first sight by analogy to the scalar case, the inverse problem should be: given R^{2T}, recover $A, B\,|_{\Omega^{c_1 T}}$. However, such a statement is not quite natural and has to be modified by the following reasons.

First, let $A = 0, B = \begin{pmatrix} b_{11} & 0 \\ 0 & b_{22} \end{pmatrix}$. In this case, the wave modes do not interact, the beam is decoupled to two independent channels with the scalar potentials b_{11} and b_{22} respectively. Correspondingly, the reply function takes the form $r = \begin{pmatrix} r_{11} & 0 \\ 0 & r_{22} \end{pmatrix}$. By the results of Part I, the function $r_{11}(t)\,|_{(0,2T)}$ determines $b_{11}\,|_{\Omega^{c_1 T}}$, whereas $r_{22}(t)\,|_{(0,2T)}$ determines $b_{22}\,|_{\Omega^{c_2 T}}$. However, such data do not determine b_{22} in $\Omega^{c_1 T} \setminus \Omega^{c_2 T}$! Hence, in the general case of the interacting channels, we should not expect that the operator R^{2T} determines A, B in $\Omega^{c_1 T} \setminus \Omega^{c_2 T}$ uniquely.

[21]The same information is obtained if one applies the controls $\begin{pmatrix} \delta \\ 0 \end{pmatrix}, \begin{pmatrix} 0 \\ \delta \end{pmatrix}$ and measures $U_x(0,t),\ 0 \le t \le 2T$.

Second, the object we are going to recover is a pair of matrix-functions A, B satisfying the self-adjointness conditions (76). Such a pair is determined by *four* independent scalar functions (parameters): the matrix elements $a_{12}, b_{11}, b_{12}, b_{22}$. In the mean time, the response operator is determined by the symmetric reply matrix-function r, which contains only *three* functional parameters r_{11}, r_{12}, r_{22} (see (112)). Therefore, to hope for the uniqueness of determination is not reasonable and we need to supplement r with 1-parameter data.

A well-motivated choice is to add the function l, which governs the slow waves, and set up the dynamical inverse problem as follows: **given the response operator R^{2T} and the delaying function $l\,|_{0<t<\left(1-\frac{c_2}{c_1}\right)T}$ of the system α^T, recover the coefficients A, B in $\Omega^{c_2 T}$.**

7.2 Solving the inverse problem

The coefficients can be recovered by the following procedure.

Step 1 Determine the constant matrices $\sqrt{\rho}$, ω, and the reply matrix function $r\,|_{(0,2T)}$ from (112). Find $a(0)$ from ω.

Find the operator C^T by (115).

Step 2 Fix $\xi \in (0, T]$. In the subspace $\mathcal{F}_l^{T,\,\xi} \subset \mathcal{F}^T$ construct a C^T-orthogonal basis of controls $\{g_k^\xi\}_{k=1}^\infty$ (see (125)).

Step 3 Compose $\left(R^T\right)^*$ by (111). Choose two constant vectors α, $\beta \in \mathbb{R}^2$.

Find the value $p(c_2\xi)$ from (135).

Step 4 Varying $\xi \in (0, T]$ and repeating the previous Steps, recover the solution p in $\Omega^{c_2 T}$. Recall that this solution is determined by the chosen vectors: $p = p_{\alpha\beta}(x)$.

Step 5 Varying properly α, β and using the equation (132) in the form

$$Ap'_{\alpha\beta} + Bp_{\alpha\beta} = p''_{\alpha\beta}$$

as a system of equations for the unknown A, B, determine these coefficients in $\Omega^{c_2 T}$ [22].

The inverse problem is solved. One more way is to make use of the GLK-equation: we can determine the control f^T from (142), recover $p(c_2\xi)$ by (136), and then continue by the Steps 4,5.

[22] Such a system determines A, B uniquely. Indeed, fix an $x_0 \in \Omega^{c_2 T}$. As the map $\{\alpha, \beta\} \mapsto p_{\alpha\beta}(x_0)$ is already recovered, one can choose nonzero α, β such that $p_{\alpha\beta}(x_0) = 0$ and then recover $A(x_0)$ from the system. Analogously, one can determine $B(x_0)$.

Remark The following additional question arises. We have recovered the coefficients in the slow interval $\Omega^{c_2 T}$. In the mean time, in accordance with the locality principle, the inverse data R^{2T}, l are determined by the behavior of the coefficients in the fast interval $\Omega^{c_1 T}$. To what extent do the inverse data determine A and B in $\Omega^{c_1 T} \setminus \Omega^{c_2 T}$? The question is answered in [12], where it is shown that R^{2T}, l do not determine $A, B\, |_{\Omega^{c_1 T} \setminus \Omega^{c_2 T}}$ uniquely and the character of non-uniqueness is revealed.

7.3 Visualization of waves

Besides the determination of coefficients, the BC-method provides a remarkable opportunity to see the waves on a beam. Of course, once the coefficients have been determined, one can find the waves just by solving the forward problem (73)–(75). However, speaking about a *visualization* we mean a certain straightforward way to recover the map $f \mapsto u^f$ through the inverse data without solving the forward problem.

Let the external observer choose a smooth control $f \in \mathcal{F}^T$. After the Steps 1,2, one can compose the operator $\mathcal{P}_l^\xi$, find the control $\mathcal{P}_l^\xi f$ (see (127)), and recover the part $u^f(\,\cdot\,, T)\,|_{\Omega^{c_2 T}}$ of the wave $u^f(\,\cdot\,, T)$ by the amplitude formula (131). Moreover, replacing in the AF the control f by the delayed one $D^{T,t} f$, the observer can determine $u^f(\,\cdot\,, t)\,|_{\Omega^{c_2 T}}$ for all $t \in (0, T]$ (see (69)) and thus visualize an *evolution* of the wave on the interval $\Omega^{c_2 T}$.
Visualization provides the picture of waves, what is much more significant than just determination of coefficients. As is easy to see, such a picture determines the coefficients. Indeed, one can visualize $u^{f_1}, u^{f_2}, \dots, u^{f_n}$ for a rich enough set of controls $f_1, f_2, \dots, f_n$ and then find A, B from the system of equations (73) written for u^{f_k}.

Visualization of waves via relevant versions of the geometrical optics formulas is the main tool for solving multidimensional inverse problems by the BC-method [4], [6], [7].

7.4 Finite beam

As we saw, in the case of the semi-infinite beam the response operator R^{2T} and the delaying function l constitute two independent parts of the inverse data. Here we show that, for a finite length beam, certain additional measurements at its second endpoint enable one to determine l and recover the coefficients everywhere on the beam.

A beam of a finite length h with the free [23] second endpoint is described

[23] i.e., not fixed and force-free

by the system

$$\rho u_{tt} - u_{xx} + A u_x + B u = 0, \qquad 0 < x < h,\ 0 < t < T \tag{143}$$

$$u|_{t=0} = u_t|_{t=0} = 0, \qquad x \geq 0 \tag{144}$$

$$u|_{x=0} = f, \quad u_x|_{x=h} = 0, \qquad 0 \leq t \leq T. \tag{145}$$

Assume that, in addition to $u^f_x(0,t)\ |_{0<t<2T}$, one can measure the values of $u^f(h,t)\ |_{0<t<T}$.

Take $T = \frac{h}{c_2}$ and choose a control $f = \begin{pmatrix} m \\ \delta \end{pmatrix}$, where $m = m(t)$ is an arbitrary smooth function. If (and only if) one puts $m = l(t)$ then the corresponding wave is slow:

$$u^{\begin{pmatrix} m \\ \delta \end{pmatrix}}\Big|_{t<\frac{x}{c_2}} = 0\,.$$

Such a wave does not manifest itself at the second endpoint and the external observer records the rest at $x = h$ for all $t \in (0,T)$. In particular, the first component of the wave, which is a scalar function[24], satisfies

$$u_1^{\begin{pmatrix} m \\ \delta \end{pmatrix}}(h,t) = 0\,, \qquad 0 < t < T\,. \tag{146}$$

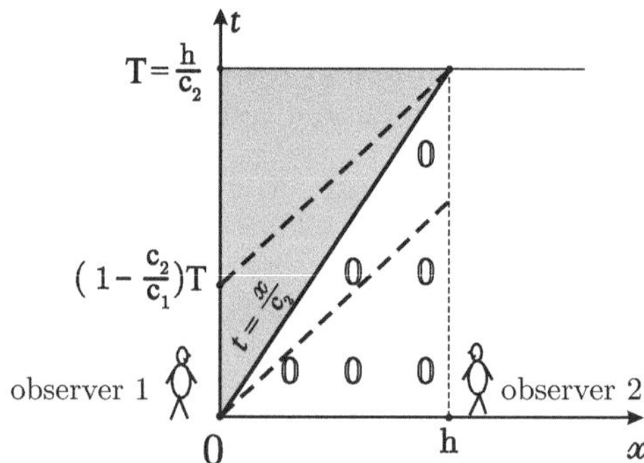

Figure 17. Observations at $x = h$

[24] This component does not contain δ-singularities: see Fig 11 c,d

It can be shown that the converse is also true: the relation (146) is valid *only if*

$$m = l(t), \qquad 0 < t < \left(1 - \frac{c_2}{c_1}\right) T$$

holds.

The latter fact provides an efficient way to find the delaying function [10], [14]. Namely, it is the function l, on which the functional

$$F[m] := \int_0^T \left[u_1^{\begin{pmatrix} m \\ \delta \end{pmatrix}}(h,t) \right]^2 dt$$

defined on smooth functions $m(t),\ 0 < t < \left(1 - \frac{c_2}{c_1}\right) T$, attains the global minimum, the minimal value being equal to $F[l] = 0$.

As soon as l is determined, one can recover the coefficients and visualize waves by means of the procedures described in the previous section.

7.5 Comments and remarks

- If $\rho_1 = \rho_2$, so that the wave modes propagate with one and the same velocity, the forward and inverse problems can be treated in the same way. Moreover, such a case is much easier for investigation. The reason is the following. Scaling the x-variable and seeking for the solution in the form $u(x,t) = S(x)v(x,t)$, one can choose the matrix-function S such that v satisfies $v_{tt} - v_{xx} + Bv = 0$ [25]. The properties of the dynamical system governed by this equation are quite analogous to ones of the scalar string: there are no slow waves, the local controllability (43) does hold. Therefore to modify the procedure of sec 4.2 for recovering the matrix-function B is a simple exercise. The theory of such systems is well known [1], [18].
- If $A = 0$ and $B = B^\sharp$, the beam equation takes the form $\rho u_{tt} - u_{xx} + Bu = 0$ and the beam is described by *three* independent functional parameters b_{11}, b_{12}, b_{22}. By this, the determination $R^{2T} \Rightarrow B$ becomes well motivated. However, in this case, our approach leads to the certain nonlinear equations [13] and it is not known, whether the suitable *linear* analog of (142) does exist. It is an interesting and important open question.

[25] i.e., one can remove the 1-order term Au_x, what is impossible if $\rho_1 \neq \rho_2$!

- In [12], the characterization of the inverse data (i.e., the necessary and sufficient conditions for R^{2T}, l to be the inverse data of a system α^T) is proposed. We have to notify that the paper contains a minor inaccuracy: the term ωf in (100) is missed and therefore the main result is valid under additional assumption $A(0) = 0$. However, to extend the characterization to the general case is not a serious task.
- One can easily modify our approach to be available for another types of the boundary conditions, e.g., for the *Neumann* control $u_x\,|_{x=0} = f$.
- Another approaches to the inverse problem for the beam based on the gradient (optimization) methods see in [27], [28]. Elementary introduction to the multidimensional BC-method can be found in [6].

Bibliography

[1] S.A.Avdonin, M.I.Belishev. Boundary control and dynamical inverse problem for nonselfadjoint Sturm-Liouville operator (BC-method). *Control and Cybernetics*, 25 (1996), No 3, 429–440.

[2] M.I.Belishev. On an approach to multidimensional inverse problems for the wave equation. *Dokl. Akad. Nauk SSSR*, 297 (1987), no 3, 524–527. English translation: *Soviet Mathematics. Doklady*, 36 (1988), no 3, 481–484.

[3] M.I.Belishev. The Gelfand–Levitan type equations in multidimensional inverse problem for the wave equation. *Zapiski Nauch. Semin. LOMI*, 165 (1987), 15–20 (in Russian); English translation: *J. Sov. Math.*, 50 (1990), no 6, 1940–1944.

[4] M.I.Belishev. Boundary control in reconstruction of manifolds and metrics (the BC method). *Inverse Problems*, 13 (1997), no 5, R1–R45.

[5] M.I.Belishev. Dynamical systems with boundary control: models and characterization of inverse data. *Inverse Problems*, 17 (2001), 659–682.

[6] M.I.Belishev. How to see waves under the Earth surface (the BC-method for geophysicists). *Ill-Posed and Inverse Problems*, S.I.Kabanikhin and V.G.Romanov (Eds). VSP, 55–72, 2002.

[7] M.I.Belishev. Recent progress in the boundary control method. *Inverse Problems*, 23 (2007), no 5, R1–R67.

[8] M.I.Belishev. Boundary control and inverse problems: 1-dimensional variant of the BC-method. *Zapiski Nauch. Semin. POMI*, 354 (2008), 19–80 (in Russian); English translation: *J. Math. Sciences*, v. 155 (2008), no 3, 343–379.

[9] M.I.Belishev, A.S.Blagoveschenskii. Dynamical Inverse Problems of Wave Theory. SPb State University, St-Petersburg, 1999 (in Russian).

[10] M.I.Belishev, A.S.Blagovestchenskii, S.A.Ivanov. Erratum to "The two-velocity dynamical system: boundary control of waves and inverse problems [Wave Motion 25 (1997) 83–107]". *Wave Motion*, 26 (1997), 99.

[11] M.I.Belishev, S.A.Ivanov. Boundary control and canonical realizations of two-velocity dynamical system. *Zapiski Nauch. Semin. POMI*, 222: 18–44, 1995 (in Russian); English translation: *J. Math. Sci.* (New York) 87 (1997), no. 5, 3788–3805

[12] M.I.Belishev, S.A.Ivanov. Characterization of data of dynamical inverse problem for two–velocity system. *Zapiski Nauch. Semin. POMI*, 259 (1999), 19–45 (in Russian); English translation: *J. Math. Sci.*, 109 (2002), no 5, 1814–1834.

[13] M.I.Belishev, S.A.Ivanov. On uniqueness "in the small" in dynamical inverse problem for a two-velocity dynamical system. *Zapiski Nauch. Semin. POMI*, 275: 41–54, 2001 (in Russian); English translation: *J. Math. Sci.* (N. Y.) 117 (2003), no. 2, 3910–3917

[14] M.I.Belishev, S.A.Ivanov. Recovering the parameters of the system of the connected beams from the dynamical boundary data. *Zapiski Nauch. Semin. POMI*, 324: 20–42, 2005 (in Russian); English translation: *J. Math. Sci.* (N. Y.) 138 (2006), no. 2, 5491–5502

[15] M.I.Belishev, T.L.Sheronova. The boundary control method in dynamical inverse problem for inhomogeneous string. *Zapiski Nauchn. Seminarov LOMI*, 186 (1990), 37-49 (in Russian). English translation: *J. Math. Sci.*, 73 (1995), no 3, 320–329.

[16] M.I.Belishev, A.V.Zurov Effects associated with the coincidence of velocities in a two-velocity dynamical system. Zapiski Nauchn. Seminarov LOMI, 264 (2000), 44–65 (in Russian). English translation: *J. Math. Sci.*, 111 (2002), no. 4, 3645–3656.

[17] A.S.Blagovestchenskii. On a local approach to the solving the dynamical inverse problem for inhomogeneous string. *Trudy MIAN V.A. Steklova* 115 (1971), 28-38 (in Russian).

[18] A.S.Blagovestchenskii. Inverse Problems of Wave Processes. *VSP, Netherlands*, 2001.

[19] I.M.Gelfand, B.M.Levitan. On the determination of a differential equation from its spectral function. *Izv. Acad. Nauk SSSR*, 15 (1951), 309-360 (in Russian).

[20] B.Gopinath and M.M.Sondhi. Determination of the shape of the human vocal tract from acoustical measurements. *Bell Syst. Tech. J.*, July 1970, 1195–1214.

[21] B.Gopinath and M.M.Sondhi. Inversion of the Telegraph Equation and the Synthesis of Nonuniform Lines. *Proceedings of the IEEE*, vol 59, no 3, March 1971, 383–392.

[22] R.Kalman, P.Falb, M.Arbib. Topics in Mathematical System Theory. *New-York: McGraw-Hill*, 1969.

[23] M.G.Krein The solving of the Sturm-Liouville inverse problem. *Dokl. Akad. Nauk SSSR*, 76 (1951), no 1, 21-24 (in Russian).

[24] M.G.Krein. On inverse problems for a non-homogeneous string. *Dokl. Akad. Nauk SSSR* 82 (1952), no 5, 669-672 (in Russian).

[25] M.G.Krein. On a method for the effective solution of the inverse boundary-value problem. *Dokl. Akad. Nauk SSSR*, 95 (1954), no 6, 767-770 (in Russian).

[26] J.-L.Lions. Contrôle optimal de syst émes gouvern 'es par les équations aux d ériv 'ees partielles. *Dunod Gauthier-Villars, Paris,* 1968.

[27] A.Morassi, G.Nakamura, M.Sini. An inverse dynamical problem for connected beams. *European J. Appl. Math*, 16 (2005), no 1, 83–109.

[28] A.Morassi, G.Nakamura, M.Sini. A variational approach for an inverse dynamical problem for composite beams. *European J. Appl. Math*, 18 (2005), no 1, 21–55.

[29] D.L.Russell. Controllability and stabilizability theory for linear partial differential equations. *SIAM Review*, 20 (1978), no 4, 639-739.

[30] V.S. Vladimirov. Equations of Mathematical Physics. *Translated from the Russian by A.Littlewood, Ed. A.Jeffrey, Pure and Applied Mathematics, 3 Marcel Dekker, Inc, New-York*, 1971.

[31] A.V.Zurov Effects associated with the coincidence of velocities in a two-velocity dynamical system. Zapiski Nauchn. Seminarov LOMI, 297 (2003), 44–65 (in Russian). English translation: *J. Math. Sci.*, 127 (2005), no. 6, 2364–2373.

Dynamic Characterization and Damage Identification

Fabrizio Vestroni and Annamaria Pau
Dipartimento di Ingegneria Strutturale e Geotecnica
Università di Roma La Sapienza, Roma, Italy

Abstract This chapter presents selected cases of modal parameter evaluation and damage identification, privileging the applied aspects. The identification of modal parameters is discussed through three experimental cases of masonry structures: a civil building, a monumental structure and a nineteenth century railway bridge. Modal parameters are response quantities that are sensitive to damage, whose variation with respect to an undamaged state can be exploited in view of damage detection and identification. Moreover, the variations of modal quantities can be used for updating the structural model, leading to the best estimate of model mechanical parameters in the undamaged and damaged conditions. This provides information on the damage location and severity. The technique of damage identification based on model updating is applied to laboratory experimental tests on a steel arch.

1 Introduction

Within the inverse problems, identification of structural models is a very meaningful subject; indeed in this book different Chapters are devoted to the matter. Notwithstanding great improvements have been made in modelling the dynamic behaviour of structural systems, due to different reasons - mainly, simplifying assumptions are generally introduced and knowledge of the actual parameter values can only be approximate - it is thoroughly felt that the experimental data obtained from the real structure can be conveniently used to make the models more reliable. The identification of models with limited initial information or realistic constraints on the unknowns to be identified, referred to as reconstruction problems, are very complex and only in few cases a unique solution is guaranteed (Gladwell, 1984; Gladwell et al., 1987; Gladwell and Morassi, 1995; Dilena and Morassi, 2010). With respect to the indeterminacy and ill-conditioning of the problems, it seems more realistic to use all the a priori information coming from the theory of materials and structures for building an initial model of the structure, then to exploit the experimental data, when available, to improve and update the model. To this goal the best use of experimental data is achieved by referring to a system identification technique that allows the suitable correlation of a priori analytical information with test results, bearing in mind that frequently little information is available from experimental tests, as few points can be instrumented, and the number of conditions that can be investigated is limited (Natke, 1982, 1991; Mottershead and Friswell, 1993; Capecchi and Vestroni, 1993, 1999; Ghanem and Sture Eds., 2000; Morassi and Vestroni, 2008).

Moreover, model identification is a fundamental step of damage evaluation and structural monitoring (Brownjohn et al., 2003; Megalhaes et al., 2008; Morassi and Vestroni, 2008). In this context a certain popularity was gained by parametric models that have a well-established structure and the only object of identification is a number of their parameters, whose optimal estimate takes advantage from the experimental information coming from the dynamic response to external excitations.

Among parametric models, here, attention is focused on two important classes: modal models and physical models, such as finite element models. Modal models are based on the assumption that the system dynamics can be represented by a series of uncoupled second-order differential equations with constant coefficients, and the modal parameters are the unknowns to be identified. In the physical models, a wider use is made of a priori information, and identification is mainly devoted to model updating by adjusting the values of the selected structural parameters. If the model is accurate in reproducing the structural behavior on the whole, these parameters will effectively maintain the physical meaning.

The choice between physical or modal models thus depends on whether it is more reasonable to formulate a realistic analytical model able to reproduce the structure dynamics with accurate values for the geometrical and mechanical characteristics, or whether it is possible to develop a sufficiently wide experimental investigation, where a high number of response quantities are measured. Notwithstanding the greater effectiveness of the physical models, which remain the final goal of the identification process, modal models are always the first step of any dynamic identification procedure, because their parameters (frequencies, eigenvectors and damping) are the basic dynamical characteristics of a structure. In particular, the experimentally evaluated modal parameters, when they are a complete set, make it possible to identify modal models that are predictive models in the direct problem, that is, they are able to furnish the response of the structure to any known excitation. When the experimental response is measured in a few points, and therefore only an incomplete set of modal eigenvalue components is determined, it is not possible to obtain an effective modal model of the structure, but the experimental values of the modal parameters determined are still important for at least two reasons: (a) their variation entails a modification of the mechanical characteristics of the structure, thus they are an efficient tool to detect damage and (b) they can be used as experimental data to identify the physical model.

Finite element models take into account all the available a priori information on the continuum models, the material behaviour and the external and internal constraints. The response to a given excitation, furnished by a finite element model, is governed by a set of coefficients that are strictly related to the physical quantities, such as, the coefficients of mass or stiffness matrices. These physical quantities represent the parameters of the models. The estimation procedure of these quantities, by using the experimental data, is generally referred to as structural identification or model updating. Uncertainties regarding system modelling are restricted to the values of some physical parameters governing the response of the model (Mottershead and Friswell, 1993). The optimal values of these are determined so as to minimize the difference between the measured and predicted response quantities. However, physical models are much more involved and, due to their theoretical structure, are more rigid with respect to modal models in fitting the given experimental results; thus, parameter identification is more difficult and onerous. Just as an example, it is difficult to reproduce the sequence of the modes with the exact ratios among the frequencies, which is an intrinsic characteristic of a structure, mainly due to the unavoidable modelling errors. This has led to a more widespread use of modal models in practical engineering, even though physical models are more attractive, as they are more robust and refined. The identification of the finite element model of a structure remains an important goal of structural identification, as it is typically used in direct problems. Moreover, it is the only model that is potentially able to give local information on the structural properties, which is crucial to damage detection (Friswell and Mottershead, 1995, 1998; Capecchi and Vestroni, 1999; Vestroni and Capecchi, 2003; Morassi and Vestroni, 2008).

Damage identification based on finite element models is a specific model updating that uses the variation of the measured data, like modal quantities, from the undamaged to damaged conditions. It is important to distinguish between two different situations: (a) diffused damage as a consequence of an infrequent event (strong wind, earthquake, explosion, etc.): this case is similar, but more complex than the initial identification of the structure, where it is reasonable to consider some regularity in the distribution of the mechanical characteristics, and the inverse problem exhibits a high number of unknowns, frequently leading to an indeterminate problem (Teughels and De Roeck, 2004; Morassi, 2007; Cruz and Salgado, 2008); (b) concentrated damage related to a local phenomenon of deterioration or overstress (corrosion, crack), in this case damage is located in a few sections and modifies few stiffness characteristics, which constitute a small number of unknowns, frequently leading to an over-determined problem (Vestroni and Capecchi, 1996; Capecchi and Vestroni, 2000; Pau et al., 2010).

In this Chapter, case-studies are presented to illustrate procedures and difficulties in modal identification and model updating for damage evaluation. As far as modal models are concerned, different methods are summarized and employed for structures excited by harmonic forces, transmitted by a mechanical shaker, and ambient excitation induced by traffic. As far as damage identification is concerned, the case of a damaged arch due to a crack is dealt with, which involves model identification in the undamaged condition and updating of the model to fit the response of the damaged structure. A procedure, which takes advantage of the peculiarity of the damage identification problem, is applied to an experimental case.

2 Identification of Modal Parameters

In structural dynamics, the multidegree-of-freedom systems represent discrete or discretized continuous systems. The governing equations of the motion of a linear system with n-degrees-of-freedom can be written in matrix form (Craig, 1981; Meirovitch, 1997):

$$\mathbf{M}\ddot{\mathbf{u}}\,(t) + \mathbf{C}\dot{\mathbf{u}}\,(t) + \mathbf{K}\,\mathbf{u}\,(t) = \mathbf{p}\,(t) \tag{1}$$

where $\mathbf{M}$, $\mathbf{C}$ and $\mathbf{K}$ are respectively the mass, damping and stiffness matrices, with dimensions $n \times n$, and $\mathbf{p}(t)$ is the $n \times 1$ force vector. If the damping matrix $\mathbf{C}$ is a combination of $\mathbf{M}$ and $\mathbf{K}$ (classical damping), the damped system will have real eigenvectors coincident with those of the undamped system, so that the equations of motion can be decoupled by describing the displacement motion in terms of the eigenvectors weighted by the modal coordinates $\mathbf{q}(t)$:

$$\mathbf{u}\,(t) = \mathbf{\Phi}\mathbf{q}\,(t) = \sum_r \mathbf{\Phi}_r \mathbf{q}_r\,(t). \tag{2}$$

By substituting this expression in the equations of motion, then premultiplying by the matrix $\mathbf{\Phi}^T$ and using the orthogonality eigenvector conditions, n uncoupled modal equations are obtained:

$$\ddot{q}_r\,(t) + 2\zeta_r\omega_r\dot{q}_r\,(t) + k_r q_r\,(t) = p_r\,(t) \qquad r = 1, \ldots n. \tag{3}$$

Hence, the system behaviour is represented as a linear superposition of the response of n single-degree-of-freedom systems. A system with n degrees of freedom has n natural frequencies ω_r and n related mode shapes $\mathbf{\Phi}_r$, along with the modal damping ratios ζ_r. These quantities are the dynamic characteristics of the structure and can be determined from the analysis of the response.

In a single-degree-of-freedom system, the Fourier transform of the response $U(\omega)$ is connected to the Fourier transform of the forcing function $F(\omega)$ by the frequency response function $H(\omega)$ (FRF), according to the linear relationship $U(\omega) = H(\omega)F(\omega)$. In n-degrees-of-freedom systems, the FRF is an $n \times n$ matrix, connecting the displacement at node i caused by a force applied at node j:

$$H_{ij}(\omega) = \frac{U_i(\omega)}{F_j(\omega)} = \sum_{r=1}^{n} \frac{\Phi_{ir}\Phi_{jr}}{k_r\left[(1-r_r^2)+i2\zeta_r r_r\right]} \tag{4}$$

where $r_r = \omega/\omega_r$ is the ratio of the forcing frequency ω to the rth system frequency ω_r. Once the registration of the response is available, there are various procedures in the time or frequency domain to determine the modal parameters of the structure (Maia and Silva, 1997). Some of these procedures require the FRFs, that is, the response to a measured input (harmonic or impulsive force) must be known. Other methods, such as the peak picking or the singular value decomposition, are based on the analysis of the output-only response. The following paragraphs simply describe these three methods, which are used in the applications presented.

2.1 Frequency Response Function Method

The method furnishes an optimal estimate of modal parameters based on the minimization of the error between the experimental FRF^e and the related analytical quantity expressed by equation (4):

$$e = H_{ij}(\omega) - H_{ij}^e(\omega). \tag{5}$$

The used algorithm is based on a multimodal estimate in the frequency domain, following the pioneering work by (Goyder, 1980). In the neighbourhood of the rth resonance, the inertance $H_{ij}(\omega)$ can be regarded as the sum of two terms:

$$H_{ij}(\omega) = \frac{-\omega^2\phi_{ir}\phi_{jr}}{-\omega_r^2-\omega^2+i2\zeta_r\omega_r\omega} + \sum_{s\neq r=1}^{N} \frac{-\omega^2\phi_{is}\phi_{js}}{\omega_s^2-\omega^2+i2\zeta_s\omega_s\omega} \tag{6}$$

The former is the prevailing resonant term and the latter is the contribution of all the other modes. By comparing the experimental and analytical inertance functions, the error at the frequency ω_k around ω_r can be defined:

$$e_k = H_{ij}(\omega_k) - H_{ij}^{(e)}(\omega_k) = \frac{-\omega_k^2 c_{ijr}}{a_r - \omega^2 + i b_r \omega_k} - C_k \tag{7}$$

where the constant C_k is given by the difference between the experimental value of the FRF and the modal contributions of all the other modes $s \neq r$, and the following modal quantities are introduced as variables of the problem:

$$a_r = \omega_r^2\,, \quad b_r = 2\zeta_r \omega_r\,, \quad c_{ijr} = \phi_{ir}\phi_{jr}. \tag{8}$$

The error function can be linearized with respect to the assumed variables by means of a suitable weight function $w_k = -1/(a_r - \omega_k^2 + i b_r \omega_k)$:

$$e_k = \frac{-\omega_k^2 \Phi_{ir}\Phi_{jr}}{\omega_r^2 - \omega_k^2 + 2i\zeta_r\omega_r\omega_k} - C_k = \omega_k^2 w_k c_{ijr} + w_k C_k \left(a_r - \omega_k^2 + i b_r \omega_k\right). \tag{9}$$

The procedure follows an iterative scheme: at the hth iteration, w_k and C_k are determined by the values at the $h-1$ step. The objective function is obtained by summing the square error in a frequency range in the neighborhood of ω_r. On the basis of the least-squares method, the modal unknowns a_r, b_r, c_{ijr} are determined. At each step, this procedure is extended to all meaningful resonances that appear in the response, and, at the end of the process, all the modal parameters in the range of frequencies investigated are determined.

2.2 Peak Picking

This method is very often used for its simplicity in analysing the ambient vibration response, when the input is unknown (Bendat and Piersol, 1980). The ambient vibration response of a structure cannot be predicted by deterministic models, within reasonable error. Each experiment produces a random time-history that represents only one physical realization of what might occur. In general, the response $x(t)$ of the structure to ambient excitation is recorded for a very long time, even for hours, which enable to cut the random process $x(t)$ into a collection of subregistrations $x_k(t)$ which describe the phenomenon.

The Fourier Transforms of the kth subregistrations of two random processes $x_k(t)$ and $y_k(t)$ are respectively:

$$X_k(f,T) = \int_0^T x_k(t) \exp^{-i2\pi ft} \, dt \tag{10}$$

$$Y_k(f,T) = \int_0^T y_k(t) \exp^{-i2\pi ft} \, dt. \tag{11}$$

The auto (or power) spectral density (PSD) and cross-spectral density (CSD) and related coherence function between the two random processes are respectively:

$$S_{xx}(f) = \lim_{T\to\infty} \frac{1}{T} E[\mid X_k(f,T) \mid^2] \tag{12}$$

$$S_{xy}(f) = \lim_{T\to\infty} \frac{1}{T} E[\bar{X}_k(f,T) Y_k(f,T)] \tag{13}$$

$$\gamma_{xy}(f) = \frac{\mid S_{xy}(f) \mid^2}{S_{xx}(f) S_{yy}(f)} \tag{14}$$

where the symbol $E[.]$ indicates an averaging operation over the index k and the overbar denotes complex conjugate.

Let us now assume that $x(t)$ is the input and $y(t)$ is the output. The auto-spectral and cross-spectral density functions satisfy the important formulae:

$$S_{yy}(f) = |H_{xy}(f)|^2 S_{xx}(f) \quad S_{xy}(f) = H_{xy}(f) S_{xx}(f) \tag{15}$$

where $H_{xy}(f)$ is the frequency response function (see equation (4)). The simple peak picking method is based on the fact that the autospectrum (15_1), at any response point, reaches a maximum either when the excitation spectrum peaks or the frequency response function peaks. To distinguish between peaks that are due to vibration modes as opposed to those in the input spectrum, a couple of criteria can be used. The former concerns the fact that in a lightly damped structure, two points must oscillate in-phase or out-of-phase. Then, the cross spectrum (15_2) between the two responses provides this information, which can be used to distinguish whether the peaks are due to vibration modes or not. The second criterion uses the coherence function (14), which tends to peak at the natural frequencies, as the signal-to-noise ratio is maximized at these frequencies.

2.3 Singular Value Decomposition

The second method referred to also relies only on the response to ambient excitations (output only). The method is based on the singular value decomposition of the response spectral matrix (Brincker et al., 2001), exploiting the relationship:

$$\mathbf{S}yy(\omega) = \bar{\mathbf{H}}(\omega)\,\mathbf{S}xx(\omega)\,\mathbf{H}^T(\omega) \tag{16}$$

where $\mathbf{S}xx(\omega)$ ($R \times R$, R number of inputs) and $\mathbf{S}yy(\omega)$ ($M \times M$, M number of measured responses) are the input and output power spectral density matrices, respectively, and $\mathbf{H}(\omega)$ is the frequency response function matrix ($M \times R$).

Supposing the inputs at the different points are completely uncorrelated and white noise, $\mathbf{S}xx$ is a constant diagonal matrix, independent of ω. Thus:

$$\mathbf{S}yy(\omega) = S\,\mathbf{H}(\omega)\,\mathbf{H}^T(\omega) \tag{17}$$

whose term jk can be written, by omitting the constant S, as:

$$Syy_{jk}(\omega) = \sum_{r=1}^{R}\left(\sum_{p=1}^{N}\frac{\phi_{jp}\phi_{rp}}{\bar{\lambda}_p^2-\omega^2}\right)\left(\sum_{q=1}^{N}\frac{\phi_{kq}\phi_{rq}}{\lambda_q^2-\omega^2}\right). \tag{18}$$

In the neighbourhood of the ith resonance, the previous equation can be approximated by:

$$Syy_{jk}(\omega) \cong \sum_{r=1}^{R}\frac{\phi_{ji}\phi_{ri}}{\bar{\lambda}_i^2-\omega^2}\frac{\phi_{ki}\phi_{ri}}{\lambda_i^2-\omega^2} = \frac{\phi_{ji}\phi_{ki}}{\left(\bar{\lambda}_i^2-\omega^2\right)\left(\lambda_i^2-\omega^2\right)}\sum_{r=1}^{R}\phi_{ri}^2. \tag{19}$$

By ignoring the constant $\sum_{r=1}^{R}\phi_{ri}^2$, $\mathbf{S}yy$ can thus be expressed as the product of the three matrices:

$$\mathbf{S}yy(\omega) = \mathbf{\Phi}\mathbf{\Lambda}_i\mathbf{\Phi}^T \tag{20}$$

which represent a singular value decomposition of the matrix $\mathbf{S}yy$, where:

$$\mathbf{\Lambda}_i = \begin{bmatrix} \frac{1}{\left(\lambda_i^2-\omega^2\right)\left(\bar{\lambda}_i^2-\omega^2\right)} & 0\ldots & 0 \\ 0 & 0\ldots & 0 \\ \vdots & \vdots & \vdots \\ 0 & 0\ldots & 0 \end{bmatrix}. \tag{21}$$

This is valid in the neighbourhood of every natural frequency of the system, that hence emerges as a peak of the first singular value. The first column of the matrix $\mathbf{\Phi}$ contains the first singular vector, which in the neighborhood of the ith resonance coincides with the ith eigenvector. This occurs at each resonance, when the prevailing contribution is given by the related mode. This procedure has recently had great diffusion mainly for *in situ* experimental tests and has also been implemented in commercial codes.

2.4 The S. Sisto School

S. Sisto school is a natural stone masonry building in L'Aquila. It consists of a main two-storey rectangular structure about 10 m high. The plan of the main building is rectangular, with sides having a length of about 12 m (Figure 1). The height of the storeys is irregular. A steel beams and hollow tiles mixed floor divides the different storeys. The building is structurally non-symmetric because of the eccentricity of the staircase. The sensors were located on the first and second floors, as shown in Figure 1, where their positions and measurement direction are indicated as (C1Y-C12X). The structure was excited with a shaker located in four different positions and directions (V1X-V4Y), also shown in Figure 1. Forced harmonic vibration tests were performed; the rotating frequency of the actuator was increased from 5 to 15 Hz with a step increment of 0.05 Hz. The modal parameters were identified by using the iterative procedure defined by the minimization of the error function (9).

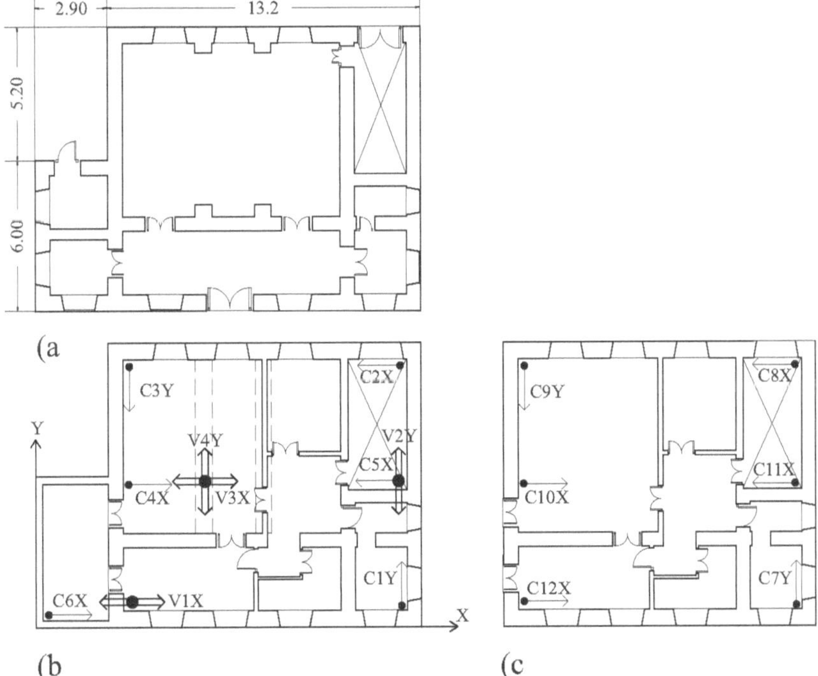

Figure 1. Building plans and locations of the accelerometers (C1Y-C12X) and of the shaker (V1X-V4Y) at ground level (a), 5.40 m (b), 9.50 m (c)

Figure 2 shows the frequency response functions (FRFs) obtained at C8X and C9Y from the four different locations of the shaker. Two groups of peaks emerge: the first represents three global modes involving the whole structure; the second contains four peaks corresponding to the higher modes, with local deformations of the structural elements. When the position of the shaker is varied, some differences in natural frequencies and mode shapes are observed. The frequencies are reported in Table 1 together with the related damping coefficients, while the mode shapes are depicted in Figure 3. Mode 1 is a bending mode along the Y axis, mode 2 is approximately a bending mode according to the X axis and mode 3 is a torsional mode. Both modes 1 and 2 involve a certain amount of torsion. This is due to the imperfect structural symmetry for the presence of the staircase.

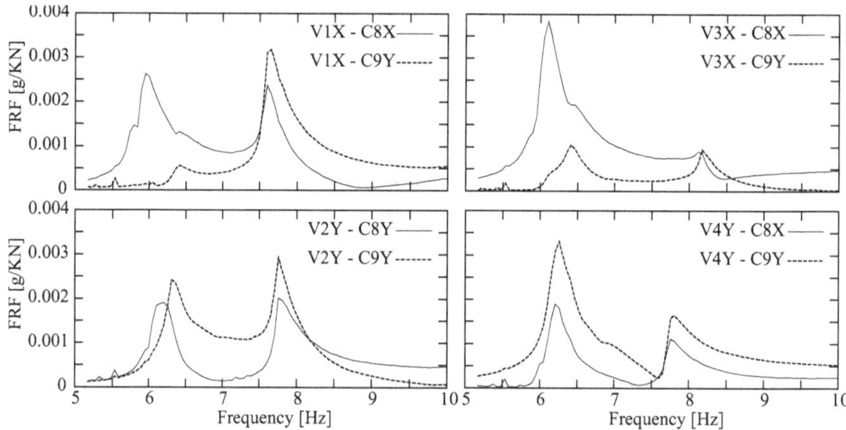

Figure 2. FRFs at C8X and C9Y for different locations of the shaker

Table 1. Natural frequencies [Hz] and damping ratios (in parentheses) obtained with the different shaker locations

Shaker location	Mode 1	Mode 2	Mode 3
V1X	5.95 (2.1%)	6.41 (2.2%)	7.65(1.6%)
V2Y	–	6.25 (2.4%)	**7.76 (1.6%)**
V3X	**6.10 (2.1%)**	6.41 (1.8%)	8.17(1.2%)
V4Y	–	**6.23 (2.2%)**	**7.76 (1.6%)**

The differences in the observed frequencies may be ascribed to nonlinearities, which often occur in masonry structures, even at a low level of vibration and to the high damping coefficient connected to the vicinity of modes. Nevertheless, a procedure for the updating of the physical model requires that only one frequency is defined for each mode, and then a selection is made by observing the time-histories on the basis of their consistency. This brings to the definition of the frequencies reported in bold in Table 1, related to V3X for the first frequency, to V4Y for the second and to V4Y and V2Y for the third. Figure 4 shows the experimental frequency response functions and those obtained by using the identified modal parameters. A satisfactory agreement is reached; the fitting is better for the first group of modes than for the second one and for the anti-resonances. The modal parameters, experimentally evaluated, are then used to update the finite element model of the structure; the details can be found in (De Sortis et al., 2005).

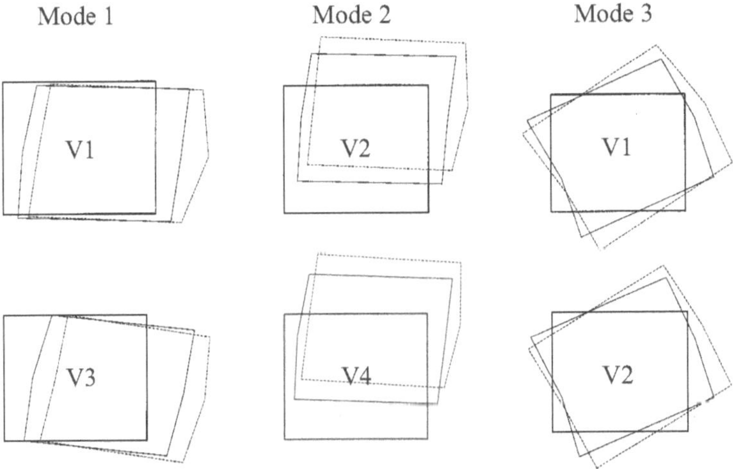

Figure 3. Comparison between the first three identified mode shapes for different shaker locations

2.5 The Flavian Amphitheatre

Known worldwide as the Colosseum, the Flavian Amphitheathre stands in the center of Rome, near to an underground line and heavy road traffic. The amphitheatre is elliptical in shape with major and minor axes being 188 m and 156 m in length, respectively; its external wall is 50 m in height.

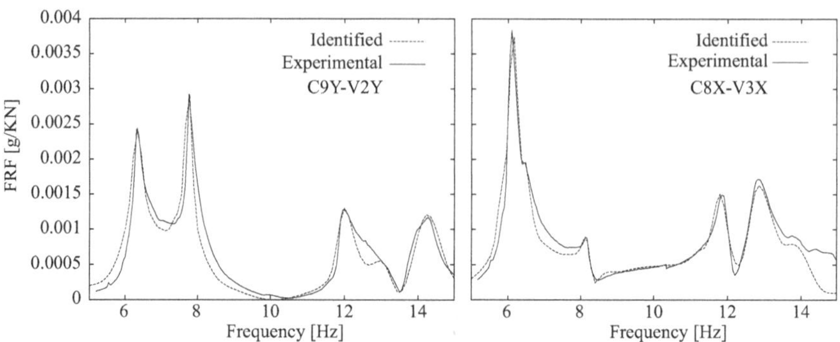

Figure 4. Comparison between experimental and identified FRFs for two sensors and shaker positions

The structure consists of a series of dry masonry travertine radial walls connected by mixed travertine/tuff concentric annular walls and by inclined concrete vaults. The heterogeneity of the soil in the foundations and various earthquakes have caused settlements that have led to collapses, which have given the Colosseum its present shape, wherein the original symmetry is partially lost.

Different kinds of tests were performed on the Flavian Amphitheater. A detailed description of all the experimental investigations can be found in (Pau and Vestroni, 2008). Triaxial recordings enabled the characterization of the excitation and vibration level. Figure 5, as an example, compares the power spectral densities (PSD) of the accelerations at the base and on the upper level of the monument (L4, about 40 m high). This figure shows that the frequency content of the vertical and tangential accelerations at the base and level L4 have similar features. By contrast, the power spectral densities of the radial acceleration reveal the filtering effect of the structure. In fact, the frequency content of the radial response at the base is similar to that in the other directions, but easily distinguishable peaks appear at the upper levels. The natural frequencies are in fact expected in this band, which is actually investigated more in detail with radial recordings. These tests have been performed by measuring accelerations along seven vertical sections with six accelerometers placed at different heights (levels L1-L6) on the external wall. When moving from one vertical to another, one extra accelerometer was kept on the previous line to track the phase. Impact vibration tests have also been carried out on the columns, using an instrumented hammer, to determine the material properties from the wave propagation velocity.

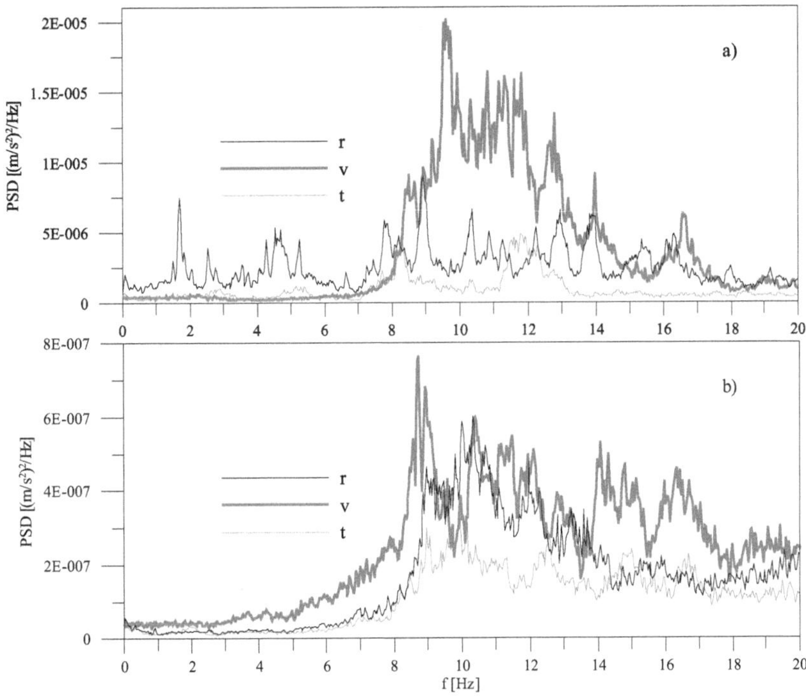

Figure 5. PSDs of accelerations at level L4 (a) and base (b)

Experimental modal analysis using ambient vibrations must face the problem of input forces being unknown. The key condition is that the excitation must have a flat spectrum in the frequency band around the frequencies to be estimated. This is verified, and then the natural frequencies are determined by simply peak picking from the PSDs. In order to obtain further evidence of the results, the coherence function γ is also observed, which also tends to peak at natural frequencies. Figure 6 (a) and (b), respectively, report the cross-spectral density (CSD) and coherence of the radial accelerations at a section. Figure 6(a) shows a couple of sharp resonance peaks at 1.03 Hz and 1.49 Hz respectively, whose amplitude increases from L1 (S_{61}) to L5 (upper level, S_{65}). At these frequencies, the coherence tends to peak, although not reaching the unit value, because of the smallness of the response amplitude (Figure 6b). Other peaks are observed in the range $1-2$ Hz: in all cases their amplitude increases from L1 to the upper levels and the coherence reaches a maximum. The values of these frequencies are reported in Table 2, which also indicates the sections where these resonance peaks are observed. Their values are summarized in the last column of the table.

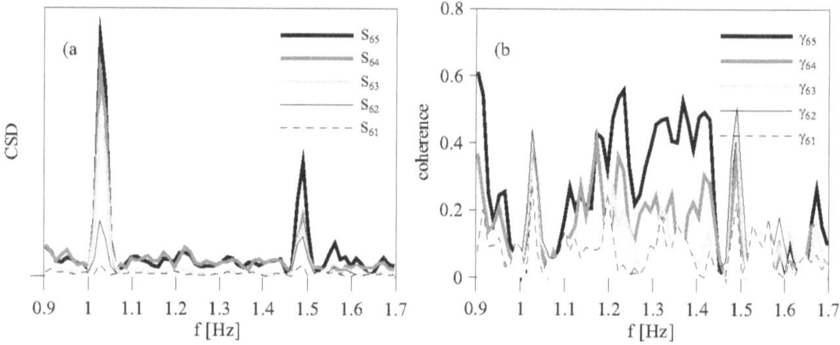

Figure 6. CSDs S_{6j} between the response at L6 and Lj (a) of radial accelerations and related coherence (b)

Table 2. First experimental frequencies [Hz] observed at different sections

section	24	27	32	39	47	51	53	f_e
L1	1.03							1.03
L2						1.30	1.30	1.30
L3	1.49	1.48	1.47		1.49	1.49		1.49
L4			1.59	1.62		1.60		1.60
L5			1.65		1.67			1.66
L6			1.75	1.73	1.76			1.75

A finite element model of the structure was also built, accurately reproducing the present geometry and using the material characteristics derived from the literature. The comparison between experimental f_e and numerical frequencies f_a and the related mode shapes is reported in Figures 7 to 9, which show that experimental frequencies are approximately half of the numerical ones. If the ratios between Young's modulus and mass density, E/ρ, of the different materials are homogeneously reduced in order to minimize the errors between f_e and f_a, these frequencies are brought into satisfactory agreement, according to Table 3. The reduction required for the ratio E/ρ is considerable, however, a similar result has already been observed in other cases of similar ancient Roman masonries. Furthermore, the updated value is strongly related to the experimental result obtained by the impact tests on the columns. In these tests, the propagation velocity of the pressure waves c_p was experimentally evaluated: the arrival time of a wave generated by an impact at one point was observed at different points and a mean experimental value of 1220 m/s was obtained for c_p. The parameter c_p is related to the mechanical properties of the medium, E, ρ and Poisson's coefficient ν according to: $c_p = \sqrt{E(1-\nu)/[\rho(1+\nu)(1-2\nu)]}$. The experimental estimate of c_p is 0.42 times the value furnished by the previous formula, using the same material properties, formerly assumed in the finite element model. This result requires a reduction in the ratio E/ρ, similar to that obtained by the process of minimisation of the differences between the natural frequencies.

Table 3. Experimental and analytical frequencies of the Colosseum [Hz] and related errors

mode	1	2	3	4	5	6
f_e	1.03	1.30	1.49	1.60	1.66	1.75
f_a	1.13	1.14	1.51	1.54	1.74	1.76
$100 \cdot \frac{\lvert f_a - f_e \rvert}{f_e}$	9.71	12.31	1.34	3.75	4.82	0.57

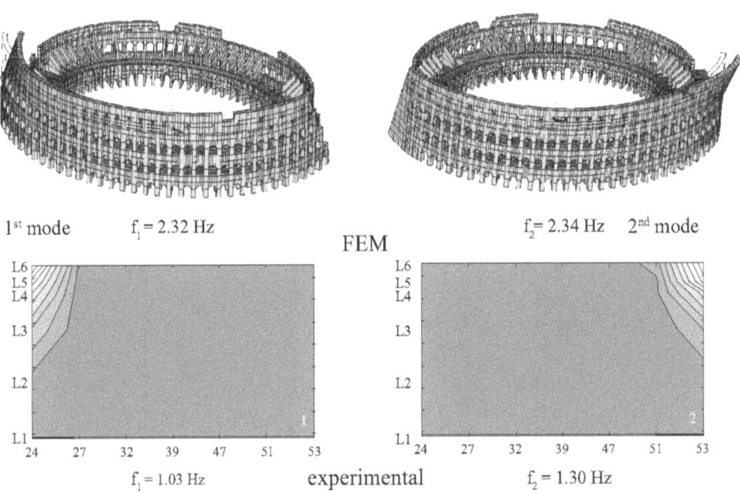

Figure 7. Comparison between 1st and 2nd analytical and experimental mode shapes

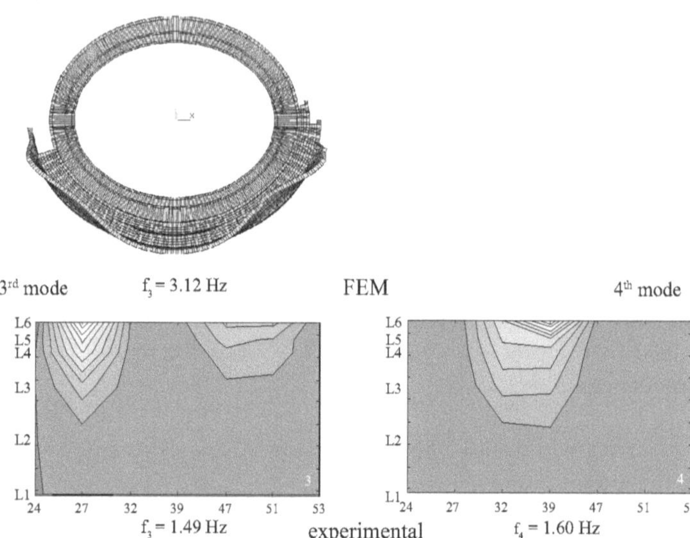

Figure 8. Comparison between 3rd analytical and experimental mode shapes and 4th experimental mode

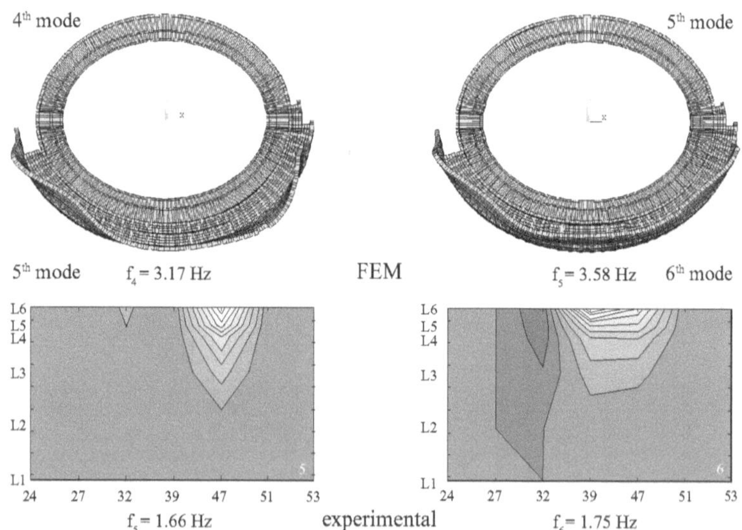

Figure 9. Comparison between 4th and 5th analytical and 5th and 6th experimental mode shapes

In order to compare the mode shapes, the values of the modal assurance criteria (MAC) between the analytical and experimental mode shapes were calculated (Table 4). The MAC is defined as the scalar product between two vectors, representing an analytical and an experimental mode, each one normalized with respect to its modulus. The first two modes are clearly identified, as appears from the high value of their MAC coefficients and also from their images in Figure 7. The experimental mode shapes are represented by the contour plots of the radial components. The accordance between the third modes is satisfactory, while the fourth experimental mode does not exhibit a similarity to any analytical mode (Figure 8). The fifth and sixth experimental modes look like the fourth and fifth analytical ones, considering that high MAC values for higher modes in the out-of-diagonal terms are mainly due to the small number of measurement points (Figure 9).

Table 4. MAC coefficients between analytical and experimental mode shapes

$a \backslash e$	1	2	3	4	5	6
1	0.93	0.10	0.27	0.03	0.01	0.03
2	0.10	0.89	0.11	0.02	0.00	0.06
3	0.14	0.12	0.69	0.23	0.49	0.51
4	0.12	0.05	0.28	0.32	0.64	0.50
5	0.13	0.02	0.69	0.20	0.42	0.62
6	0.15	0.09	0.37	0.47	0.45	0.39

2.6 The Vallone Scarpa Bridge

The Vallone Scarpa bridge was built at the end of the nineteenth century and is located along the Roma-Sulmona railway line, which crosses the central Italian region of Abruzzo. It is a masonry arch viaduct with thirteen bays, each with a span of 10 m. The piers are about 9 m in height. The plan has a radius of curvature of 400 m; the slope of the line is 2.7 %.

The ambient vibrations of the bridge were recorded using two different arrangements of accelerometers. The measurement directions were: in the plan of the deck, tangent to the bridge axis (L, longitudinal) and related orthogonal line (T, transverse); and the vertical direction (V) along the viaduct axis. In the first setup, transverse sensors were placed on the top of each pier, together with biaxial sets (T, L) located at the middle of each span. In the second setup, triaxial sets of accelerometers (L,T,V) were placed on the deck edges of the three central bays.

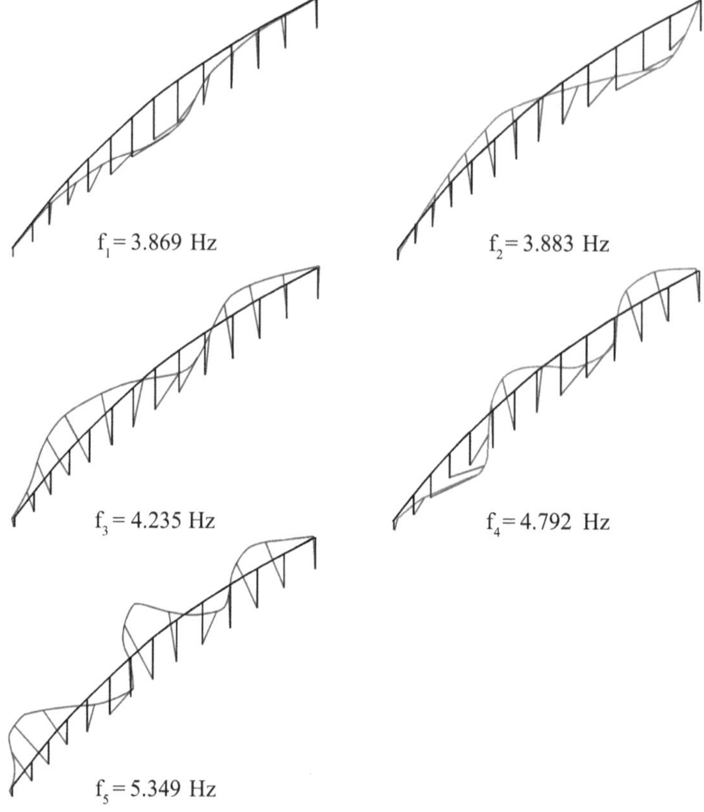

Figure 10. Experimental frequencies and mode shapes of the Vallone Scarpa bridge.

The natural frequencies and mode shapes were determined from the ambient vibration responses by using the previously described technique of singular value decomposition. Using the first arrangement of sensors, the first five frequencies were determined, showing that the mode shapes are bending modes of a beam over elastic supports in the plane of the deck, as shown in Figure 10, which depicts the first five modes of the structure and reports the related frequencies. The main component of the mode shape is transverse, as shown in Figure 11, which reports, for the sake of brevity, a comparison between the longitudinal and transverse components of the first and second mode shapes only. The measurements performed with the second arrangement of sensors provided similar results, but showed in addition that the identified modes present a slight rotational component, which is pointed out by the fact that the deck edges oscillate out-of-phase. This is shown in Figure 12, only for modes 1 and 2 for brevity. Furthermore, in the second arrangement of sensors there was a slight variation of frequencies, but an inversion between the order of modes 1 and 2 was found (see Table 5). According to these measurements, the first mode presents one node, while the second one does not have any. This phenomenon is related to the closeness between the two frequencies of the arch in the horizontal plane and to the possible slight variation of mechanical parameters. A complete explanation of the phenomenon would require the repetition of the measurements and verification of their robustness with regard to the ambient conditions.

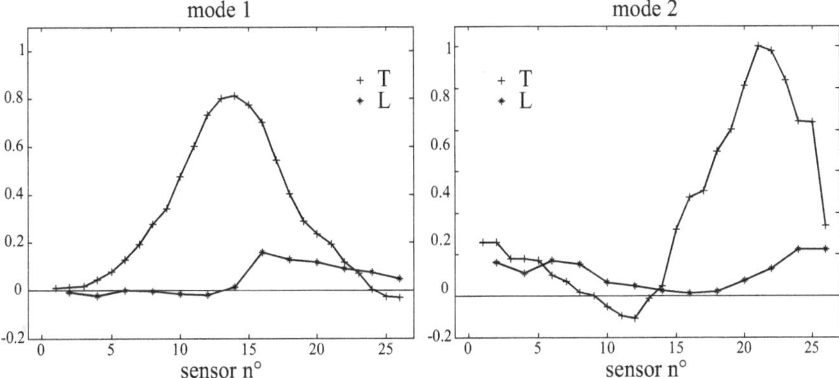

Figure 11. Comparison between longitudinal and transverse components of the first and second experimental modes.

3 Damage Identification

In structural identification problems, the system model is characterized by the structural parameters x, which are to be identified by the measured output z^* to the input I^*. The system with generic parameters x, excited by the true input I^*, does not reproduce the exact output z^*, but responds with the generic output z. The challenge consists in identifying the correct values of the parameters x^*, so as to give the response values close to the measured ones.

An important application of the identification techniques is damage evaluation. When dealing with damaged structures, a modification of the re-

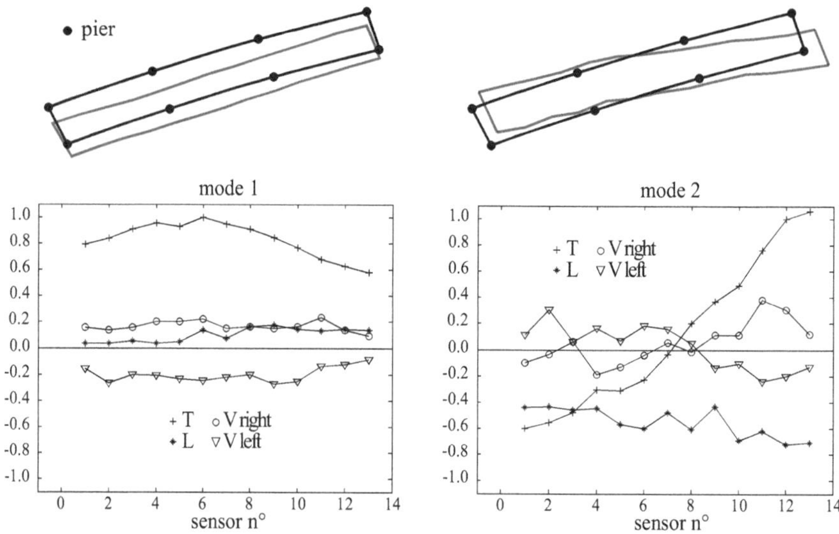

Figure 12. Comparison between longitudinal, transverse and vertical components of experimental modes 1 and 2.

Table 5. Frequencies of the Vallone Scarpa bridge determined with the two sensor arrangements

mode	1	2	3	4	5
first setup	3.869	3.883	4.235	4.792	5.349
second setup	3.854	3.839	4.191	4.909	5.510

sponse due to damage is observed. The same input I^* acting on the undamaged structure with parameters x^U produces an output z^U which is different from the output z^D of the damaged structure with parameters x^D. Variations in the behaviour of a structure can be associated with the decay of the system mechanical properties (Shen and Pierre, 1990; Morassi, 1993). The response in a damaged state can be represented as a deviation from the undamaged state $z^U + \Delta z^D$ due to a variation of the parameters $x^U + \Delta x^D$.

Two possible approaches can be followed: in the first approach, the model parameters are identified in the two different states, the undamaged (x^U) and the damaged (x^D) states; the variation in the parameters $(x^U - x^D)/x^U$ furnishes the degree of damage. In the second approach, using a given model of the structure, the modification $\Delta x^D/x^U$ are determined in such a manner as to fit, at best, the experimental modification of the observed quantities Δz^D.

Both problems can be formulated as minimization problems of suitable functionals $\mathfrak{L}$, which in the absence of errors, is equal to zero for the exact values $(.)^*$ and are greater than zero for values different from the exact ones $(.)^*$:

$$\mathfrak{L}(x^*; z^*) = 0 \quad \text{and} \quad \mathfrak{L}(x; z^*) > 0 \quad \text{for} \quad x \neq x^* \tag{22}$$

$$\mathfrak{L}(\Delta x^*; \Delta z^*) = 0 \quad \text{and} \quad \mathfrak{L}(\Delta x; \Delta z^*) > 0 \quad \text{for} \quad \Delta x \neq \Delta x^*. \tag{23}$$

Damage identification in a one-dimensional continuum due to open cracks is an interesting problem in practice and simple enough for discussing the basic aspects. A reduction in stiffness is interpreted as damage.

In the past, this problem has been tackled as a general identification problem: the conditions for spectrum reconstruction have been discussed. The variations in the dynamic response due to cracks has been examined in numerous articles, a consistent number of which are specifically directed at predicting the location and magnitude of the cracks. Among all these studies, little attention has been paid to some peculiar aspects of damage identification. It is important to distinguish between two different situations: (a) diffused damage, where the problem exhibits a high number of unknowns, it is certainly indeterminate and additional conditions must be considered; (b) concentrated damage, when damage is located in few sections; in this case, damage modifies few stiffness characteristics and the problem has a small number of unknowns. Unfortunately, in both cases, the problem has always been tackled by assuming all parameters to be unknown. Hence, the result is indeterminate and conditions not generally related to the physical behaviour have been introduced to make it determinate (Vestroni and Capecchi, 1996; Cerri and Vestroni, 2000; Morassi and Vestroni, 2008).

3.1 Damage Identification in a Steel Arch

An experimental application of a damage identification procedure in a steel arch with concentrated damage, consisting of a notch on a given cross-section is presented. The damage parameters to be identified are the location s and stiffness constant k of a spring equivalent to the notch for its effects on the local increase in deformability. Four damage configurations were considered, characterized by the same location (s=0.3) with increasing intensity (D1-D4); the expected values of k are reported in the first column of Table 6. The procedure is based on the comparison between numerical and experimental variations of natural frequencies from undamaged to damaged states; hence the reference (undamaged) state has to be known.

In the solution to the inverse problem, it is important to be aware of the necessary number of measured quantities providing a unique solution. This goal can be reached by studying some aspects of the direct problem through a model, analytical or numerical, of the arch. In particular, for a damaged configuration characterized by the values of k and s, the model provides the values of the natural frequencies ω_i. On the contrary, when ω_i is known for the damaged arch, for each possible damage position s, a stiffness $k_i(s)$ exists, which corresponds to a value of the ith natural frequency equal to ω_i. Bearing in mind more complex situations, a FE model is adopted to solve the eigenvalue problem of the arch. Therefore, by taking into account all the possible positions of the damage, a curve $k_i(s_j)$ can be obtained. Figure 13 represents the curves $k_i(s_j)$ for the first five frequencies of vibration, obtained by using as damage parameters $s=$ 0.3 and $k=$ 20. This figure can be used to examine the uniqueness of the solution to the inverse problem. Curves $k_i(s_j)$ obtained for different ω_i cross at the abscissa where the damage is located, providing the solution to the inverse problem. In fact, only at that point the values of k and s satisfy all the assigned frequencies. In the case of simply supported straight beams, the curves corresponding to the first and the second frequencies, $k_1(s_j)$ and $k_2(s_j)$ cross only once, providing the solution to the inverse problem by using only two frequencies (Figure 13 b). Regarding the arch presented here, however, each pair of curves shows more than one intersection, and thus the solution is undetermined when only two frequencies are known (Figure 13a). This is due to the wavelengths of curves $k_i(s_j)$, which present at least one node along the length of the arch, one more than the simply supported straight beam. Thus, at least three different frequencies are necessary to uniquely determine the two damage parameters.

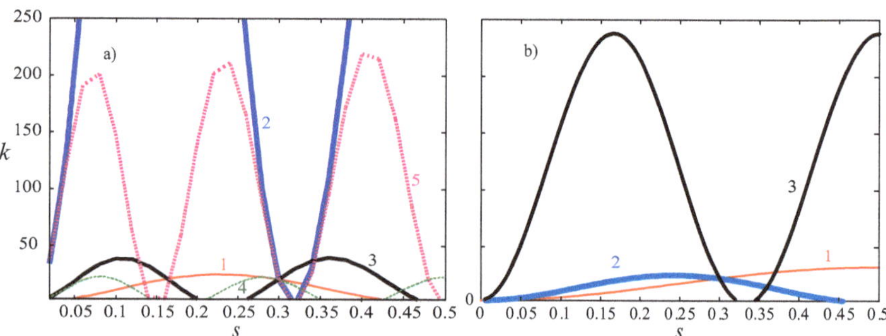

Figure 13. Curves $k_i(s_j)$ for the first five modes of the arch a) and three modes of a beam b); $s=$ 0.30, $k=$ 20.

It should also be observed that the axis of the figures reported involves only one half of the beam because of the symmetry of the problem, which does not enable one to discern between two symmetric damage positions. This is a drawback of the frequency approach, which could be overcome by adding some information on the mode shapes or on anti-resonant data, as shown in (Dilena and Morassi, 2010).

In real cases, when errors are present and more measurements than those strictly required are available, it is convenient to formulate the problem as a minimization problem. The optimal estimate of the damage parameters $\overline{k}$ and $\overline{s}$ is obtained by minimizing the objective function:

$$G(k, s) = \sum_{i} \left(\frac{\Delta\omega_i(k, s)}{\omega_i^U} - \frac{\Delta\omega_{ei}}{\omega_{ei}^U} \right)^2 . \tag{24}$$

This is defined as the sum of the squares of the differences between the numerical $\Delta\omega_i(k, s)$ and experimental $\Delta\omega_{ei}$ variations of frequencies between the undamaged and the damaged states, normalized with respect to the frequencies of the undamaged arch, ω_i^U and ω_{ei}^U.

As the numerical values $\Delta\omega_i(k, s)$ are provided by a FE model of the arch, it is convenient to obtain the damage parameters $\overline{k}$ and $\overline{s}$ in two phases, by successively seeking two distinct minima. For each possible discrete damage position s_j in the FE model, the minimisation of function (24) with respect to k provides the function:

$$\widetilde{G}(s_j) = \min_k G(k, s_j). \tag{25}$$

The solution to the inverse problem is then given by the minimum of $\widetilde{G}(s_j)$ over s. If this function exhibits one global minimum, the solution to the inverse problem exists and is unique.

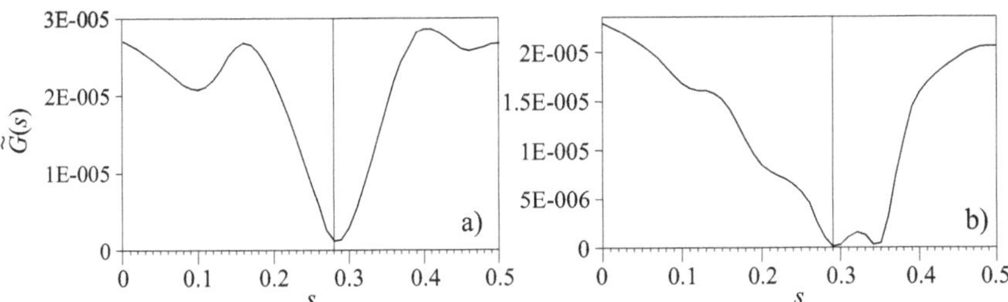

Figure 14. Objective functions using frequencies 1-5 a) and 1,2,5 b) for the case D2.

As observed in the direct problem, the minimum number of frequencies necessary to have a unique solution is three. This is further confirmed by Figure 14, which reports the objective functions built with five (a) and three frequencies (b), and in particular frequencies 1-2-5, related to the experimental case of damage D2. For brevity, only case D2 is reported. For both functions (a) and (b), the global minimum is unique, notwithstanding the fact that other local minima appear, due to close multiple intersections in the curves $k_i(s_j)$. The use of more than three frequencies does not necessarily provide better results, as is shown in Table 6, referring to the objective function with five frequencies and Table 7 for three frequencies. These tables report the values of the spring stiffness and damage positions obtained at the global minimum of the objective function for the four damage configurations considered. Both objective functions provide notable errors only in the weakest damage scenario (D1); in all other damage scenarios satisfactory results are obtained. The error strongly decreases with increasing damage, where systematic errors are less important. However, case (b) provides, in general, smaller errors due to the exclusion of outliers (third frequency) or frequencies having a low sensitivity to damage in the considered position (fourth frequency).

Table 6. Expected and identified stiffness, notch depths δ, damage locations and errors (a, i=1-5)

	$\bar{k}$	k_{id}	% error	δ[mm]	% error	s	% error
D1	570	1161	103.7	0.6	44.3	0.74	3.6
D2	205	249	21.7	1.8	11.1	0.72	0.8
D3	91	98	8.2	2.9	3.3	0.71	0.0
D4	40	36	10.7	4.1	3.3	0.70	1.4

Table 7. Identified stiffness, notch depths, damage locations δ and related errors using frequencies 1,2,5 (b)

	k_{id}	% error	δ[mm]	% error	s	% error
D1	527	7.5	1.1	6.2	0.65	8.5
D2	205	0.2	2.0	0.1	0.71	0.0
D3	89	1.4	3.0	0.6	0.72	0.8
D4	36	10.1	4.1	3.1	0.71	0.0

In the weakest damage scenario (D1), the frequency variations are small, and the accuracy of the model becomes a key factor. In fact, in this case, the addition of an axial spring modelling the axial stiffness reduction helps

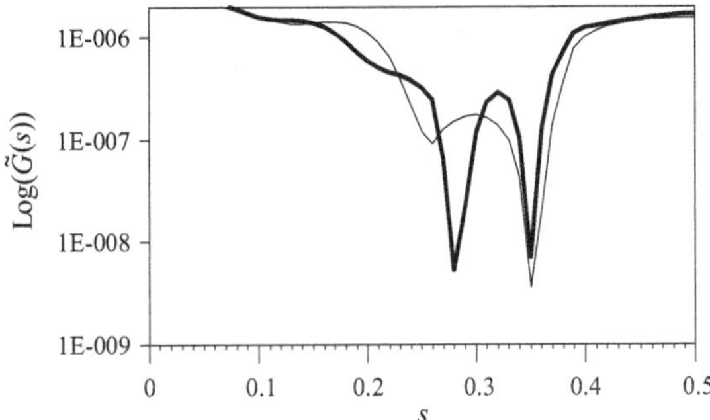

Figure 15. Objective functions obtained from the model with (bold) and without (thin) reduction of axial stiffness for the case D1.

in identifying the right notch location. This is shown in Figure 15, which shows the objective functions obtained when using three frequencies and wether the reduction in the axial stiffness induced by damage is considered (bold line) or not. More details can be found in (Pau et al., 2010).

4 Conclusions

There is a growing interest in the experimental assessment of the models of existing structures, to better understand and predict their behaviour and, more importantly, to know how their evolution may be used within a health monitoring strategy. Within the inverse problems, attention has been focused here only on parametric model identification, where available experimental data are used to obtain an optimal estimate of the model parameters. In particular two categories of models are considered: modal models and finite element models. In the first the unknowns to identify are the modal parameters (frequencies, eigenvector components, damping), in the second the parameters are the mechanical characteristics of the model.

Parametric identification is a typical inverse problem, and as such it is generally a difficult task; however, modal model identification is easier and this explains its wider use and the implementation of the procedure in commercial codes. The most important methods are summarized and case-studies have been presented for two different scenarios, when input and output are measured and when output-only is measured; this latter case is rapidly spreading, because nowadays the recording of ambient vibrations is feasible even for large structures. The results presented for large and complex structures, such as historical buildings, are a demonstration that experimental modal analysis is reliable enough to furnish an experimental evaluation of the modal parameters and a dynamic characterization of the structure; the evolution of the parameters makes it possible to find out the modification of the structural characteristics and to detect damage. Modal parameters are still the experimental quantities most frequently used in the successive identification of more refined models. The identification of the physical parameters of the finite element models of a structure is more involved, but these parameters are representative of local structural characteristics; their evolution is potentially able to give a detailed description of damage. The identification of damage in an arch due to an open crack has been presented to illustrate the use of FE model updating, based on the experimental data, like modal quantities, evaluated in undamaged and damaged conditions. The procedure, which takes into account the peculiarity of this inverse problem, is able to localize damage and to determine its intensity, a step more ambitious than damage detection; the results are satisfactory, but it has been shown how important the accuracy of the model and quality of experimental data are.

Bibliography

J.S. Bendat and A.G. Piersol. *Engineering Applications of Correlation and Spectral Analysis.* John Wiley & Sons, New York, 1980.

R. Brincker, L. Zhang, and P. Andersen. Modal identification of output-only systems using frequency domain decomposition. *Smart Materials and Structures*, 10:441–445, 2001.

J.M.W. Brownjohn, P. Moyo, P. Omenzetter, and Y. Lu. Assessment of highway bridge upgrading by dynamic testing and finite element model updating. *Journal of Bridge Engineering*, 8:162–172, 2003.

D. Capecchi and F. Vestroni. Damage detection in beam structures based on measurements of natural frequencies. *Journal of Engineering Mechanics ASCE*, 126(7):761–768, 2000.

D. Capecchi and F. Vestroni. Monitoring of structural systems by using frequency data. *Earthquake Engineering and Structural Dynamics*, 28: 447–461, 1999.

D. Capecchi and F. Vestroni. Identification of finite element models in structural dynamics. *Engineering Structures*, 15(1):21–30, 1993.

M. N. Cerri and F. Vestroni. Detection of damage in beams subjected to diffused cracking. *Journal of Sound and Vibration*, 234(2):259–276, 2000.

R.R. Craig. *Structural Dynamics.* John Wiley & Sons, New York, 1981.

P.J.S. Cruz and R. Salgado. Performance of vibration-based damage detection methods in bridges. *Computer-Aided Civil and Infrastructure Engineering*, 24:62–79, 2008.

A. De Sortis, E. Antonacci, and F. Vestroni. Dynamic identification of a masonry building using forced vibration tests. *Engineering Structures*, 27:155–165, 2005.

M. Dilena and A. Morassi. Reconstruction method for damage detection in beams based on natural frequency and antiresonant frequency measurements. *Journal of Engineering Mechanics*, 136(3):329–344, 2010.

M.I. Friswell and J.D. Mottershead. *Finite Element Model Updating in Structural Dynamics*. Kluwer Academic Publishers, Dordrecht, 1995.

M.I. Friswell and J.D. Mottershead. Identification in engineering systems, special issue. *Journal of Vibration and Control*, 4(1), 1998.

R. Ghanem and S. Sture Eds. *Journal of Engineering Mechanics*, 126(7): 665–777, 2000.

G.M.L. Gladwell. The inverse problem for the vibrating beam. *Proceedings of the Royal Society of London Series A*, 393:277–295, 1984.

G.M.L. Gladwell and A. Morassi. On isospectral rods, horns and string. *Inverse Problems*, 11(3):533–554, 1995.

G.M.L. Gladwell, A.H. England, and D. Wang. Examples of reconstruction of an Euler-Bernoulli beam from spectral data. *Journal of Sound and Vibration*, 119:81–94, 1987.

H.G.D. Goyder. Methods and applications of structural modelling from measured frequency response data. *Journal of Sound and Vibration*, 68: 209–230, 1980.

N.M.M. Maia and J.M.N. Silva. *Theoretical and Experimental Modal Analysis*. Reaserch Studies Press, Baldock, 1997.

F. Megalhaes, A. Cunha, and E. Caetano. Dynamic monitoring of a long span arch bridge. *Engineering Structures*, 30:3034–3044, 2008.

L. Meirovitch. *Principles and Techniques of Vibration*. Prentice-Hall, Upper Saddle River, N. J., 1997.

A. Morassi. Crack-induced changes in eigenparameters of beam structures. *Journal of Engineering Mechanics*, 119(9):1798–1803, 1993.

A. Morassi. Damage detection and generalized Fourier coefficients. *Journal of Sound and Vibration*, 302(1–2):229–259, 2007.

A. Morassi and F. Vestroni. *Dynamic Methods for Damage Detection in Structures*. Springer, Wien, 2008.

J. E. Mottershead and M. I. Friswell. Model updating in structural dynamics: a survey. *Journal of Sound and Vibration*, 167(2):347–375, 1993.

H.G. Natke. *Identification of vibrating structures*. Springer–Verlag, New York, 1982.

H.G. Natke. *in: Structural Dynamics*. Recent trends in system identification. W.B. Kratzing et al Eds Springer–Verlag, Balkema, Netherlands, 1991.

A. Pau and F. Vestroni. Vibration analysis and dynamic characterization of the Colosseum. *Structural Control & Health Monitoring*, 15:160–165, 2008.

A. Pau, A. Greco, and F. Vestroni. Numerical and experimental detection of concentrated damage in a parabolic arch by measured frequency variations. *Journal of Vibration and Control*, pages 1–11, 2010.

S.H. Shen and C. Pierre. Natural modes of Bernoulli-Euler beams with symmetric cracks. *Journal of Sound and Vibration*, 138(1):115–134, 1990.

A. Teughels and G. De Roeck. Structural damage identification of the highway bridge Z24 by FE updating. *Journal of Sound and Vibration*, 278:589–610, 2004.

F. Vestroni and D. Capecchi. *Parametric identification and damage detection in structural dynamics*, pages 107–143. Research Signpost, Trivandrum, India, 2003. A. Luongo ed.

F. Vestroni and D. Capecchi. Damage evaluation in cracked vibrating beams using experimental frequencies and finite element models. *Journal of Vibration and Control*, 2:69–86, 1996.

Eigenvalue Assignment Problems in Vibration Using Measured Receptances: Passive Modification and Active Control

John E Mottershead*, Maryam G Tehrani* and Yitshak M Ram†

* Department of Engineering,
University of Liverpool, United Kingdom

† Department of Mechanical Engineering
Louisiana State University, Baton Rouge, USA

Abstract Matrix methods using mass, damping and stiffness terms are widely used in vibration analysis of structure and provide the basis for active control of vibrations using state-space methods. However modern vibration test procedures provide reliable receptance data with usually much greater accuracy than is achievable from system matrices, **M**, **C**, **K**, formed from finite elements. In this article procedures are described for pole placement by passive modification and active control using measured receptances. The theoretical basis of the method is described and experimental implementation is explained.

1 Introduction

Receptances were first used in the passive modification of structures by Duncan (1941) and the pole assigment problem was described as a classical inverse problem in active control by Kautsky et al. (1985). In this article the use of measured receptances for pole assignment in passive modification and active control is described. The considerable advantages of this approach are explained.

In conventional modelling the dynamic behaviour of a structure is determined from mass, damping and stiffness matrices, **M**, **C**, **K**, usually obtained from finite elements. In theory the receptance matrix, which is measurable experimentally, is the inverse of the dynamic stiffness matrix,

$$\mathbf{H}(s) = \mathbf{Z}^{-1}(s) \tag{1}$$

$$\mathbf{Z}(s) = s^2\mathbf{M} + s\mathbf{C} + \mathbf{K} \tag{2}$$

In reality finite element models contain approximations, assumptions and errors not present in measured data. The terms in the receptance matrix, $\mathbf{H}(s)$, are dominated by the eigenvalues (or poles) closest to the frequency of excitation,

$$\mathbf{H}(s) = \sum_{k=1}^{n} (\frac{\varphi_k \varphi_k^T}{s - \lambda_k} + \frac{\varphi_k^* \varphi_k^{*T}}{s - \lambda_k^*}) \tag{3}$$

where φ_k denotes the kth eigenvector.

It may be shown, however, that the dynamic stiffness terms, in $\mathbf{Z}(s)$, are dominated by the poles furthest away - usually the higher modes, most affected by discretisation and the least accurate from a finite element model. We see that $\mathbf{H}(s)$ may be represented very accurately by a truncated set of modes whereas $\mathbf{Z}(s)$ generally may not be. Furthermore,the problem of dynamic-stiffness error cannot be overcome by inverting the accurately measured matrix of receptances. Within the limited frequency range of a vibration test, and when $s = \mathtt{i}\omega$, the inversion of $\mathbf{H}(\mathtt{i}\omega)$ is not able to reproduce $\mathbf{Z}(\mathtt{i}\omega)$ because the effect of the high-frequency modes is negligibly small, within the measurement noise (Berman and Flannelly, 1971).

The receptance terms may be expressed by the ratio of two polynomials,

$$\mathbf{H}(s) = \frac{\mathbf{N}(s)}{d(s)} \tag{4}$$

where $d(\lambda_k) = 0$ is the characteristic equation of the original system, with eigenvalues λ_k, $k = 1, 2, \ldots, 2n$. The ij^{th} term in $\mathbf{N}(s)$, provides the characteristic equation of the zeros, $n_{ij}(\xi_k) = 0$, $k = 1, 2, \ldots, 2m$, $m \leq n$. This means that the vibration response at coordinate i vanishes when a force is applied at coordinate j with frequency ξ_k. The assignment of zeros is important for delicate instruments etc. that will not tolerate harsh vibration environments.

The polynomial $d(s)$ and the matrix $\mathbf{N}(s)$ may be expanded as,

$$d(s) = \prod_{k=1}^{n} ((s - \lambda_k)(s - \lambda_k^*)) \tag{5}$$

and

$$\mathbf{N}(s) = \begin{bmatrix} \Phi & \Phi^* \end{bmatrix} \mathtt{adj} \begin{bmatrix} \mathtt{diag}(s - \lambda_j) & 0 \\ 0 & \mathtt{diag}(s - \lambda_j^*) \end{bmatrix} \begin{bmatrix} \Phi & \Phi^* \end{bmatrix}^T \tag{6}$$

where Φ is the modal matrix of eigenvector columns. It follows from equation (6) that $\texttt{rank}(\mathbf{N}(s)) = 1$ if and only if $s = \lambda_k;\ \lambda_k \in \lambda_j;\ j = 1, 2, \ldots, n,$

$$\mathbf{N}\left(\lambda_k\right) = \varphi_k \left(\prod_{j=1, j \neq k}^{n} \left(\lambda_k - \lambda_j\right) \prod_{j=1}^{n} \left(\lambda_k - \lambda_j^*\right) \right) \varphi_k^T \tag{7}$$

The dynamic stiffness equation,

$$\mathbf{Z}(s)\mathbf{x}(s) = \mathbf{f}(s) \tag{8}$$

is a 'force' equation. Every term in $\mathbf{x}(s)$, the displacement or state vector, must be present for the equation to be complete. This is the reason why, in state-space control problems, it is necessary to use an observer (or equivalent) to estimate the unmeasured states. In industrial-scale problems the number of coordinates can be many thousands, or even millions, and therefore it is necessary to use model reduction methods to condense the problem to an acceptable size. On the other hand, the receptance equation, expressed as

$$\mathbf{H}(s)\mathbf{f}(s) = \mathbf{x}(s) \tag{9}$$

is a 'displacement' equation. For each row of the matrix equation to be complete it is only necessary to know the applied forces, usually at a small number of coordinates. Therefore for every sensor location, the measured terms in $\mathbf{x}(s)$, one only needs to know the non-zero terms in $\mathbf{f}(s)$, which are usually measured in an experiment. Therefore, there is no requirement for an observer or for model reduction when using the receptance method.

In practical problems we may be interested to modify the spectrum (the natural frequencies and damping ratios) of a system, usually to avoid resonances with excitation frequencies. This may be achieved by modifying the system, either passively by adding mass, damping or stiffness terms, or by active control. The advantage of passive modification is that the modified system is guaranteed to remain stable. However, as will be seen later, practical modification requires difficult measurements, typically rotational receptances. The advantage of active control is that it offers much greater freedom in the 'form' of the modification, but this is achieved at the expense of 'compensation' to ensure stability. The simplest modification is the *unit-rank modification* and a straightforward solution is available by the Sherman-Morrison formula,

$$\overline{\mathbf{H}}(s) = \mathbf{H}(s) - \frac{\mathbf{H}(s)\mathbf{u}(s)\mathbf{v}^T(s)\mathbf{H}(s)}{1 + \mathbf{v}^T(s)\mathbf{H}(s)\mathbf{u}(s)} \tag{10}$$

where $\overline{\mathbf{H}}(s)$ is the receptance matrix of the modified system. The characteristic equation is,

$$1 + \mathbf{v}^T(\mu_k)\mathbf{H}(\mu_k)\mathbf{u}(\mu_k) = 0 \tag{11}$$

where μ_k, $k = 1, 2, \ldots, 2n$ are the eigenvalues of the modified system.

2 Passive Modification

Probably the simplest passive modification is a spring connecting a single coordinate to earth. In this case,

$$\mathbf{u} = \mathbf{v} = \mathbf{e}_p \tag{12}$$

where $\mathbf{e}_p$ is the pth unit vector and p denotes the spring attachment coordinate. The characteristic equation becomes,

$$1 + \tilde{k}\mathbf{e}_p^T \mathbf{H}(\mu_k)\mathbf{e}_p = 0 \tag{13}$$

The spring $\tilde{k}$ that will exactly assign the eigenvalue μ_k is given by,

$$-\frac{1}{\tilde{k}} = h_{pp}(\mu_k) \tag{14}$$

where $h_{pp}(\mu_k) = \mathbf{e}_p^T \mathbf{H}(\mu_k)\mathbf{e}_p$, the pth diagonal receptance term. A single added spring can assign only one eigenvalue exactly and the other eigenvalues are then given by solving the characteristic equation (13). The modified stiffness matrix is given by $\mathbf{K} + \tilde{k}\mathbf{e}_p\mathbf{e}_p^T$.

Example 1. We consider a conservative lumped-mass, five degree of freedom system, where,

$$\mathbf{K} = \begin{bmatrix} 2 & -1 & & & \\ -1 & 2 & -1 & & \\ & -1 & 3 & -1 & \\ & & -1 & 2 & -1 \\ & & & -1 & 2 \end{bmatrix}; \; \mathbf{M} = \mathbf{I}$$

The eigenvalues and eigenvectors are given by,

$$\Lambda = \begin{bmatrix} 0.7203 & & & & \\ & 1 & & & \\ & & 1.5202 & & \\ & & & 1.7321 & \\ & & & & 2.0421 \end{bmatrix}$$

$$\Phi = \begin{bmatrix} 0.3578 & 0.5 & 0.5765 & -0.5 & 0.1993 \\ 0.5299 & 0.5 & -0.1793 & 0.5 & -0.4325 \\ 0.4271 & 0.0 & -0.5207 & 0.0 & 0.7392 \\ 0.5299 & -0.5 & -0.1793 & -0.5 & -0.4325 \\ 0.3578 & -0.5 & 0.5765 & 0.5 & 0.1993 \end{bmatrix}$$

It is found from equation (14) that a spring of stiffness $\tilde{k} = 1.19$ at coordinate 3 will assign an eigenvalue at 1.6 rad/s. In Figure 1 we see that the intersections of the horizontal line at $-1/\tilde{k}$ with h_{33} (denoted by the dotted line) coincide with the peaks of $\bar{h}_{33}$. The eigenvalues at 1.0 and 1.7321 rad/s are uncontrollable by the added spring, which is attached at a node, and therefore remain unchanged. They are also unobservable and connot be seen in Figure 1.

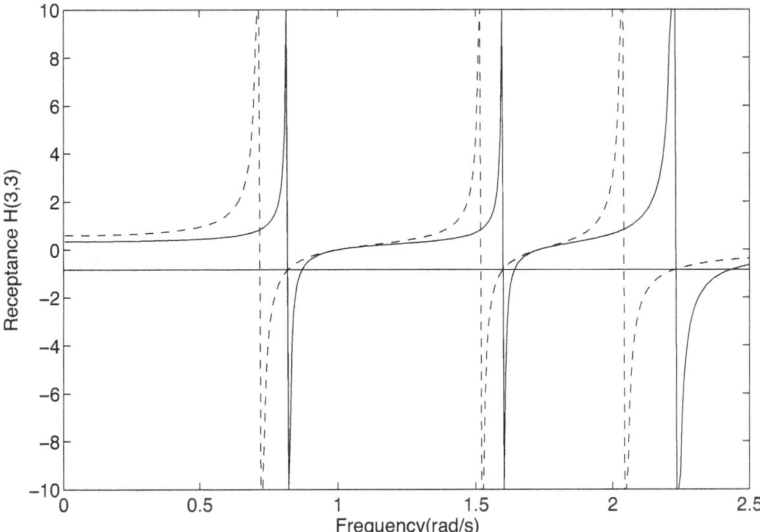

Figure 1: Unmodified and modified receptances

For further detailed information on the assignment of poles, zeros and vibration nodes by rank-1 passive modification the reader is referred to Mottershead and Ram (2006), Tehrani et al. (2006) and Mottershead et al. (2001).

2.1 Passive Modifications of Higher Rank

Most practical passive modifications are not rank-1, but of considerably higher rank. A practical example might be to connect a beam between two points on a structure. If the connection points are unconstrained then there are generally six degrees of freedom, three translations and three rotations, that must be connected. This usually requires a modification of at least rank-6. Likewise, if the modification takes the form of a large mass, there will be both translational and rotational inertia amounting to six degrees of freedom to be connected. The modified receptance matrix may be written as (Kyprianou et al., 2005),

$$\bar{\mathbf{H}}(s) = (\mathbf{I} + \mathbf{H}(s)\Delta\mathbf{Z}(s))^{-1}\mathbf{H}(s) \tag{15}$$

where

$$\Delta\mathbf{Z}(s) = s^2\Delta\mathbf{M} + s\Delta\mathbf{C} + \Delta\mathbf{K} \tag{16}$$

denotes the dynamic stiffness of the modification. The characteristic equation of the poles becomes,

$$\mathtt{det}(\mathbf{I} + \mathbf{H}(\mu_k)\Delta\mathbf{Z}(\mu_k)) = 0 \tag{17}$$

What appears, from the mathematical formulation, to be an essentially straightforward problem of pole assignment is in fact very difficult in practice. The difficulty lies in the measurement of the receptance matrix, $\mathbf{H}(s)$. In fact, receptances are not measured directly but are estimated from multiple sets of accelerometer and force-sensor measurements. The estimation may be arranged to minimise either the noise on the force signal or the noise on the output signal. The latter, known as the H_1 estimate is probably the most widely used. Translational accelerometers and actuators have been in use for many decades and there is no difficulty in estimating point- and cross-receptances between translational inputs and outputs. The problem arises with the need to estimate those receptances that involve a rotational input or output. For example to estimate a rotation-rotation receptance it is necessary to impart a pure moment (without a force) to the structure and also to measure the rotational acceleration. Although rotational accelerometers are now commercially available it appears that the quality of measurement is not as good as conventional translational accelerometers. One way to overcome the problem is to excite the structure with forces and moments together and then to measure all the resulting translations and rotations; this becomes a multiple-input multiple-output problem. Mottershead et al. (2005) and Mottershead et al. (2006) used an array of accelerometers on a ‘T’ or ‘X’-block attachment and by moving an electromagnetic shaker to different points on the attachment in turn, excitation was applied to the structure in many degrees of freedom. Of course, the attachment introduced a certain amount of additional mass to the system and was itself flexible. The arms of the ‘T’ or the ‘X’ (shown in Figures 2(a) and 2(b)) need to be sufficiently long that a significant moment excitation can be achieved. Also the effects of the forces applied at the tips of the arms need to be different and this is only possible if the tips are separated by a sufficient distance.

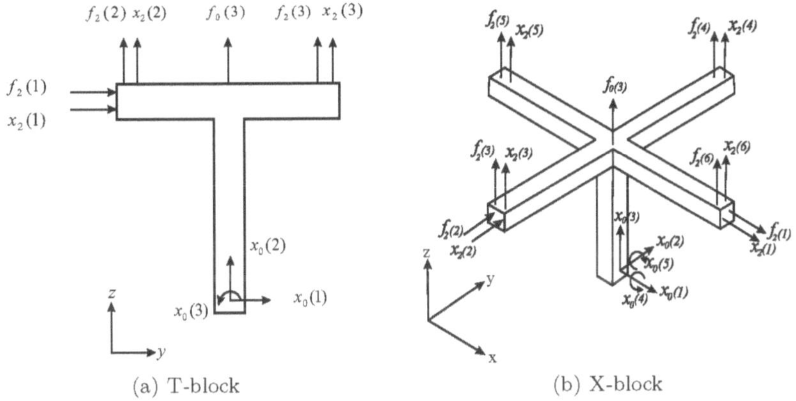

(a) T-block (b) X-block

Figure 2: (a)T-block and (b)X-block forcing and response coordinates

This requirement leads to an additional complication; that the attachment is likely to have natural frequencies of its own within the range of the test. This was overcome (Mottershead et al., 2005), (Mottershead et al., 2006) by embedding a finite element model of the attachment into the estimator for the matrix of receptances.

2.2 Modification by an Added Beam or a Large Mass

The Γ-structure modified by an added beam to form a portal frame is shown in Figure 3. The 'T'-block attachment is shown in Figure 4. The inverse problem was to assign natural frequencies using equation (17). A full description may be found in Mottershead et al. (2005). The dynamic stiffness matrix, $\Delta \mathbf{Z}(s)$ was parameterised using the cross-sectional dimensions of the added rectangular hollow-section beam, $b = \texttt{breadth}, d = \texttt{depth}, t = \texttt{wall thickness}$.

The characteristic equations, being high-order multivariable polynomials in b, d and t, were solved using the method of Groebner bases. In the cases of assigning a single natural frequency, two parameters were held constant and solutions for the third (variable) parameter were found. When two natural frequencies, or a natural frequency and an antiresonance, were to be assigned, then two parameters were varied. Among the multiple solutions produced, a value was found in each case corresponding (closely) to the parameter value in a beam used in the experiment to complete the portal frame. Typical results, presented in Table 1, show complex numbers, with small imaginary parts, returned for the variable parameters. This is because

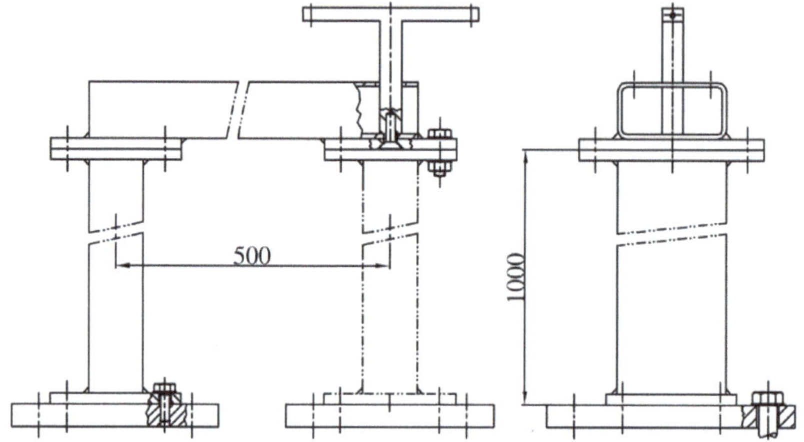

Figure 3: Portal frame structure with a removable beam

Figure 4. T-block experiment

the Γ-structure contains a small amount of damping, which needs to be compensated in order to place the poles (natural frequencies) exactly on the imaginary axis of the complex s-plane.

Table 1: Assigning (i) a natural frequency at 44.25Hz, (ii)two natural frequencies at 44.25Hz and 89.57Hz and (iii) a natural frequency at 44.25Hz and an antiresonance at 89.97Hz.

	Parameters	Fixed parameters	Solution
(i)	t	$(b,d)=(0.1,0.05)$	0.00406 + 0.00627i
(ii)	(b,d)	$t=0.004$	0.0924−0.0008i,0.0516+0.0022i
(iii)	(b,d)	$t=0.004$	0.0924−0.0008i,0.0516+0.0022i

In a different experiment (Mottershead et al., 2006) the matrix of receptances from a Westland Helicopters Lynx Mark 7 was estimated using an 'X'-block attachment. The effect of adding a large overhanging mass (equivalent to the tail rotor gearbox and hub) was then determined. The added mass was equal to the mass of the complete tailcone and therefore comprised a very significant modification. The problem, of course, was a forward one rather than an inverse problem. Figures 5 and 6 show the tailcone set-up including the use of the 'X'-attachment (Figure 5) and the added mass (Figure 6). The measurement of rotational receptances requires the use of special techniques not available from commercial modal test systems, which can be very time consuming, and a very high level of experimental expertise to obtain good quality results. However, as we see in the following section, the requirement of specialist techniques and expertise may be avoided completely when poles and zeros are assigned by active vibration control.

3 Active Control

The advantages to be had by pole placement using active vibration control are very considerable. Firstly there is no restriction to the form of the modification. Whereas passive modifications must be symmetric, positive-definite and conform to a certain pattern of non-zero matrix terms, no such restrictions apply in active control. Secondly, there is no requirement for rotational receptances. Finally, the number of eigenvalues to be assigned may be much greater than the number of actuators. Whereas in passive modification the number of eigenvalues to be assigned cannot exceed the rank of the modification, in active control a single actuator can in principle assign all $2n$ poles by a rank-1 modification.

Figure 5: X-block attachment to the tailcone

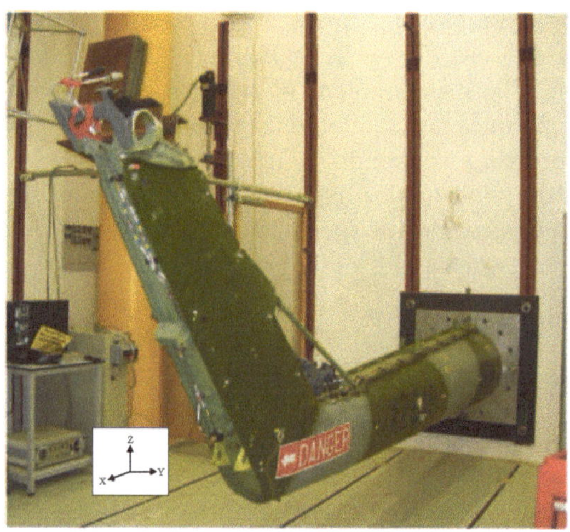

Figure 6: Lynx Mark 7 helicopter tailcone with the added mass

We begin by considering the case of single-input state feedback, and write the second order matrix equation as,

$$(s^2\mathbf{M} + s\mathbf{C} + \mathbf{K})\mathbf{x} = \mathbf{b}u(s) + \mathbf{p}(s) \tag{18}$$

or

$$\left(s\begin{bmatrix} \mathbf{0} & \mathbf{M} \\ \mathbf{M} & \mathbf{C} \end{bmatrix} + \begin{bmatrix} -\mathbf{M} & \mathbf{0} \\ \mathbf{0} & \mathbf{K} \end{bmatrix}\right)\begin{pmatrix} s\mathbf{x} \\ \mathbf{x} \end{pmatrix} = \begin{pmatrix} \mathbf{0} \\ \mathbf{b} \end{pmatrix} u(s) + \mathbf{p}(s) \tag{19}$$

where

$$u(s) = -(s\mathbf{f} + \mathbf{g})^T \mathbf{x} \tag{20}$$

is the control input and $\mathbf{p}(s)$ is an external disturbance.

Orthogonality conditions for the quadratic pencil were developed by Fawzy and Bishop (1976) and more recently by Datta et al. (1997),

$$\varphi_k^T \left[(\lambda_j + \lambda_k)\mathbf{M} + \mathbf{C}\right]\varphi_j = 0 \tag{21}$$

$$\varphi_k^T \left[\lambda_j\lambda_k\mathbf{M} - \mathbf{K}\right]\varphi_j = 0 \tag{22}$$

$$j \neq k$$

and

$$\varphi_k^T \left[2\lambda_k\mathbf{M} + \mathbf{C}\right]\varphi_k = 1 \tag{23}$$

$$\varphi_k^T \left[\lambda_k^2\mathbf{M} - \mathbf{K}\right]\varphi_k = \lambda_k \tag{24}$$

By combining equations (18) and (20) we see that,

$$\left(s^2\mathbf{M} + s\left(\mathbf{C} + \mathbf{b}\mathbf{f}^T\right) + \left(\mathbf{K} + \mathbf{b}\mathbf{g}^T\right)\right)\mathbf{x}(s) = \mathbf{p}(s) \tag{25}$$

amounts to a rank-1 modification to the dynamic stiffness matrix.

Now, after omitting the external disturbance $\mathbf{p}(s)$, setting $s = \lambda_k$ and premultiplying equation (19) by $\left(\lambda_k\varphi_k^T \quad \varphi_k^T\right)$,

$$\begin{pmatrix} \lambda_k\varphi_k^T & \varphi_k^T \end{pmatrix}\left(\lambda_k\begin{bmatrix} \mathbf{0} & \mathbf{M} \\ \mathbf{M} & \mathbf{C} \end{bmatrix} + \begin{bmatrix} -\mathbf{M} & \mathbf{0} \\ \mathbf{0} & \mathbf{K} \end{bmatrix}\right)\begin{pmatrix} \lambda_k\varphi_k \\ \varphi_k \end{pmatrix}$$

$$= -\left(\varphi_k^T\mathbf{b}\right)\left(\left(\mathbf{g}^T + \lambda_k\mathbf{f}^T\right)\varphi_k\right) \tag{26}$$

The right-hand-side vanishes whenever $\left(\varphi_k^T\mathbf{b}\right) = 0$, the *uncontrollability* condition, or $\left(\mathbf{g}^T + \lambda_k\mathbf{f}^T\right)\varphi_k = 0$, the *unobservability* condition. We observe that under either of these conditions the eigenvalue λ_k remains unchanged by control action.

3.1 Single-Input State Feedback by the Receptance Method

Application of the Sherman-Morrison formula to the rank-1 modification described in equation (25) leads to the modified receptance matrix (Ram and Mottershead, 2007),

$$\overline{\mathbf{H}}(s) = \mathbf{H}(s) - \frac{\mathbf{H}(s)\mathbf{b}\,(\mathbf{g} + s\mathbf{f})^T\,\mathbf{H}(s)}{1 + (\mathbf{g} + s\mathbf{f})^T\,\mathbf{H}(s)\mathbf{b}} \tag{27}$$

The closed-loop characteristic equation may be written as,

$$1 + (\mathbf{g} + \mu_k\mathbf{f})^T\,\mathbf{H}(\mu_k)\mathbf{b} = 0 \tag{28}$$

or,

$$d(\mu_k) + (\mathbf{g} + \mu_k\mathbf{f})^T\,\mathbf{N}(\mu_k)\mathbf{b} = 0 \tag{29}$$

From equations (29) and (7) we see that the system is uncontrollable (unobservable) when $\mathbf{N}(\mu_k)\mathbf{b} = 0$ $((\mathbf{g} + \mu_k\mathbf{f})^T\,\mathbf{N}(\mu_k) = 0)$; $\mu_k = \lambda_k$.

The characteristic equations for each assigned eigenvalue may now be assembled together in matrix form as,

$$\begin{bmatrix} \mathbf{r}_1^T & \mu_1\mathbf{r}_1^T \\ \mathbf{r}_2^T & \mu_1\mathbf{r}_2^T \\ \vdots & \vdots \\ \mathbf{r}_{2n}^T & \mu_1\mathbf{r}_{2n}^T \end{bmatrix} \begin{pmatrix} \mathbf{g} \\ \mathbf{f} \end{pmatrix} = \begin{pmatrix} -1 \\ -1 \\ \vdots \\ -1 \end{pmatrix} \tag{30}$$

or,

$$\mathbf{G}\begin{pmatrix} \mathbf{g} \\ \mathbf{f} \end{pmatrix} = \begin{pmatrix} -1 \\ -1 \\ \vdots \\ -1 \end{pmatrix} \tag{31}$$

where,

$$\mathbf{r}_k = \mathbf{H}\left(\mu_k\right)\mathbf{b} \tag{32}$$

It may be seen that $\mathbf{G}$ is invertible when the closed-loop eigenvalues to be assigned are simple and controllable. In principle the complete spectrum of $2n$ eigenvalues may be assigned using a single actuator (i.e., a single non-zero term in $\mathbf{b}$).

Example 2. We consider the system described by,

$$\mathbf{M} = \begin{bmatrix} 1 & 0 \\ 0 & 2 \end{bmatrix};\ \mathbf{C} = \begin{bmatrix} 1 & -1 \\ -1 & 1 \end{bmatrix};\ \mathbf{K} = \begin{bmatrix} 8 & -3 \\ -3 & 3 \end{bmatrix};\ \mathbf{b} = \begin{bmatrix} 1 \\ 2 \end{bmatrix}$$

The poles to be assigned are,

$$\mu_1 = -1 + 10\text{i};\ \mu_2 = -1 - 10\text{i};\ \mu_3 = -2;\ \mu_1 = -3$$

Then, we find that $\mathbf{G}$ is given by,

$$\begin{bmatrix} 0.0102 + 0.0021\text{i} & -0.0097 + 0.0020\text{i} & -0.0110 - 0.1043\text{i} & -0.0101 - 0.0989\text{i} \\ 0.0102 - 0.0021\text{i} & -0.0097 - 0.0020\text{i} & -0.0110 + 0.1043\text{i} & -0.0101 + 0.0989\text{i} \\ 0.1236 & 0.2360 & -0.2472 & -0.4719 \\ 0.0714 & 0.1111 & -0.2143 & -0.3333 \end{bmatrix}$$

The control gains are found to be,

$$\mathbf{g} = \begin{bmatrix} 68.8750 \\ 30.3750 \end{bmatrix};\ \mathbf{f} = \begin{bmatrix} -62.6750 \\ 68.1750 \end{bmatrix}$$

It is readily demonstrated that the desired eigenvalues are assigned, by solving the eigenvalue problem,

$$\left(\mu_k \begin{bmatrix} \mathbf{0} & \mathbf{M} \\ \mathbf{M} & \mathbf{C} + \mathbf{b}\mathbf{f}^T \end{bmatrix} + \begin{bmatrix} -\mathbf{M} & \mathbf{0} \\ \mathbf{0} & \mathbf{K} + \mathbf{b}\mathbf{g}^T \end{bmatrix}\right) \begin{pmatrix} \mu_k \varphi \\ \varphi \end{pmatrix} = \mathbf{0}$$

The method described above was extended by Ram et al. (2009) to include the effects of time delay such as occur in analogue to digital (A to D) and D to A conversions, use of digital filters, other digital processes and slow response of actuators etc. Mottershead et al. (2008) considered the problem of pole-zero placement using output feedback with collocated actuators and sensors. This results in a higher rank modification, greater than rank-1.

3.2 Partial Pole Placement

The purpose of partial pole placement is to assign certain chosen poles (usually the lower frequency poles) while other poles (usually at higher frequencies) remain unchanged. The usual justification given for doing this is the avoidance of 'spillover'. Generally, when a fewer number of poles than the full $2n$ are assigned the remaining ones are changed and may then have positive real parts causing the system to be unstable. It is this change to the unassigned poles that is referred to as 'spillover'. Therefore by avoiding spillover completely using partial pole placement, there is no possibility that the closed-loop system will become unstable. We should qualify this by saying that partial pole placement should be applied robustly to minimise the effects of uncertainty in the control gains and receptance measurements.

There are two ways of formulating the partial pole placement problem using either the conditions of unobservability (Datta et al., 1997) or uncontrollability (Tehrani et al., 2010). We begin by partitioning the spectrum and the modes. The poles to be assigned are denoted by,

$$\begin{bmatrix} \Lambda_1 & \\ & \Lambda_1^* \end{bmatrix} = \begin{bmatrix} \mathtt{diag}(\mu_j) & \\ & \mathtt{diag}(\mu_j^*) \end{bmatrix}; j = 1, 2, \ldots, m \tag{33}$$

and those to be retained by,

$$\begin{bmatrix} \Lambda_2 & \\ & \Lambda_2^* \end{bmatrix} = \begin{bmatrix} \mathtt{diag}(\lambda_j) & \\ & \mathtt{diag}(\lambda_j^*) \end{bmatrix}; j = m+1, m+2, \ldots, n \tag{34}$$

The corresponding modes are denoted by,

$$\begin{bmatrix} \Psi_1 & \Psi_1^* \end{bmatrix} = \begin{bmatrix} \psi_1 & \cdots & \psi_m & \psi_1^* & \cdots & \psi_m^* \end{bmatrix} \tag{35}$$

$$\begin{bmatrix} \Psi_2 & \Psi_2^* \end{bmatrix} = \begin{bmatrix} \psi_{m+1} & \cdots & \psi_n & \psi_{m+1}^* & \cdots & \psi_n^* \end{bmatrix} \tag{36}$$

Datta et al. (1997) showed that for an arbitrary vector β and,

$$\mathbf{f} = -\mathbf{M} \begin{bmatrix} \Psi_1 & \Psi_1^* \end{bmatrix} \begin{bmatrix} \Lambda_1 & \\ & \Lambda_1^* \end{bmatrix} \beta; \quad \mathbf{g} = \mathbf{K} \begin{bmatrix} \Psi_1 & \Psi_1^* \end{bmatrix} \beta \tag{37}$$

that the poles $\begin{bmatrix} \Lambda_2 & \\ & \Lambda_2^* \end{bmatrix}$ remained unchanged. From the unobservability condition, $\left(\mathbf{g}^T + \lambda_k \mathbf{f}^T\right) \psi_k = 0$, and using equations (37),

$$\beta^T \left(- \begin{bmatrix} \Lambda_1 & \\ & \Lambda_1^* \end{bmatrix} \begin{bmatrix} \Psi_1 & \Psi_1^* \end{bmatrix}^T \mathbf{M} \psi_k \lambda_k + \begin{bmatrix} \Psi_1 & \Psi_1^* \end{bmatrix}^T \mathbf{K} \psi_k \right) = 0 \tag{38}$$

Since $\lambda_k \in \Lambda_2$; $\psi_k \in \Psi_2$ it becomes clear (from orthogonality) that the bracketed term on the left-hand-side is the null vector. The unobservability condition is fulfilled so that $\mu_k = \lambda_k$ remains unchanged irrespective of the choice of β. An explicit formula for β is proveded by Datta et al. (1997).

The advantage of this approach is that there is no need to know the spectrum and modes of the unchanged eigenvalues. However, use of the orthogonality condition has the important practical implication that a sensor must be located at every degree of freedom of the system. It is not necessary for there to be an actuator at every degree of freedom. The vector β may be chosen freely to assign the closed loop poles μ_k, μ_k^*, $k = 1, \ldots, m$ while independently the other open-loop poles μ_k, $\mu_k^* = \lambda_k$, λ_k^*, $k = m+1, \ldots, n$ are retained.

The other way to achieve partial pole placement is by using the uncontrollability condition, $\psi_k^T \mathbf{b} = 0$. We now express $\mathbf{b}$ in the form,

$$\mathbf{b}(s) = \mathbf{b}_1 + \frac{\mathbf{b}_2}{s} \tag{39}$$

since $\mathbf{b}$ must be perpendicular to both the real and imaginary parts of ψ_k. Of course there are other acceptable forms of $\mathbf{b}$ apart from equation (39). Then for the retained eigenvalues $\mathbf{b}(s)$ is sought that satisfies the expression,

$$\begin{bmatrix} \Psi_2^T & \Lambda_2^{-1}\Psi_2^T \end{bmatrix} \begin{pmatrix} \mathbf{b}_1 \\ \mathbf{b}_2 \end{pmatrix} = \mathbf{0} \tag{40}$$

or,

$$\begin{pmatrix} \mathbf{b}_1 \\ \mathbf{b}_2 \end{pmatrix} = \mathbf{V}\alpha \tag{41}$$

$$\mathbf{V} = \texttt{null}\begin{bmatrix} \Psi_2^T & \Lambda_2^{-1}\Psi_2^T \end{bmatrix} = \texttt{null}\begin{bmatrix} \Re\left(\Psi_2^T\right) & \Re\left(\Lambda_2^{-1}\Psi_2^T\right) \\ \Im\left(\Psi_2^T\right) & \Im\left(\Lambda_2^{-1}\Psi_2^T\right) \end{bmatrix} \tag{42}$$

and α is an arbitrarily chosen vector. The gains $\mathbf{g}$ and $\mathbf{f}$ may then be selected to assign the closed-loop poles μ_k, μ_k^*, $k = 1, \ldots, m$.

The advantages of this approach are firstly that the spectrum may be separated into assigned and retained eigenvalues without using the orthogonality conditions. This means that there is no requirement to know or to evaluate the system matrices $\mathbf{M}$, $\mathbf{C}$, $\mathbf{K}$, which is entirely consistent with the receptance-based approach. Secondly it is not necessary to place sensors at all the degrees of freedom. It is however necessary to know the retained modes at the chosen sensor coordinates, but this is not usually a major problem when using a modern modal test system. In principle, the number of actuators required is one greater than the number of retained modes.

Example 3. We consider the system described by,

$$\mathbf{M} = \begin{bmatrix} 3 & & & \\ & 10 & & \\ & & 20 & \\ & & & 12 \end{bmatrix}; \ \mathbf{C} = \begin{bmatrix} 2.3 & -1 & & \\ -1 & 2.2 & -1.2 & \\ & -1.2 & 2.7 & -1.5 \\ & & -1.5 & 1.5 \end{bmatrix}$$

$$\mathbf{K} = \begin{bmatrix} 40 & -30 & & \\ -30 & 60 & -30 & \\ & -30 & 90 & -30 \\ & & -30 & 30 \end{bmatrix}$$

The open-loop poles are,

$$\begin{aligned} \lambda_{1,2} &= -0.0108 \pm 0.8736\mathtt{i} \\ \lambda_{3,4} &= -0.809 \pm 1.6766\mathtt{i} \\ \lambda_{5,6} &= -0.1336 \pm 2.5280\mathtt{i} \\ \lambda_{7,8} &= -0.3980 \pm 4.0208\mathtt{i} \end{aligned}$$

We wish to assign the first two pairs of poles,

$$\mu_{1,2} = -0.03 \pm 1\mathtt{i}$$
$$\mu_{3,4} = -0.1 \pm 2\mathtt{i}$$

while the remaining poles are unchanged.
The eigenvectors of the unassigned poles are found to be,

$$\psi_{5,6} = \begin{bmatrix} -0.0941 \mp 0.2578\mathtt{i} \\ -0.0829 \mp 0.1727\mathtt{i} \\ 0.1056 \pm 0.2807\mathtt{i} \\ -0.0738 \mp 0.1775\mathtt{i} \end{bmatrix}; \ \psi_{7,8} = \begin{bmatrix} -0.0535 \pm 0.2107\mathtt{i} \\ 0.0220 \mp 0.0613\mathtt{i} \\ -0.0033 \pm 0.0077\mathtt{i} \\ 0.0006 \mp 0.0014\mathtt{i} \end{bmatrix}$$

Then,

$$\mathtt{null}\left(\begin{array}{c} \Re\begin{pmatrix} \psi_5^T & \frac{\psi_5^T}{\lambda_5} \\ \psi_7^T & \frac{\psi_7^T}{\lambda_7} \end{pmatrix} \\ \Im\begin{pmatrix} \psi_5^T & \frac{\psi_5^T}{\lambda_5} \\ \psi_7^T & \frac{\psi_7^T}{\lambda_7} \end{pmatrix} \end{array} \right)$$

$$= \begin{bmatrix} 0.1996 & -0.0023 & -0.0649 & 0.0436 \\ 0.7146 & 0.0394 & -0.2611 & 0.1725 \\ 0.2723 & 0.3580 & -0.4771 & 0.3078 \\ -0.5712 & 0.5065 & -0.4113 & 0.2447 \\ 0.1143 & 0.1676 & 0.0730 & -0.0610 \\ 0.1512 & 0.6275 & 0.3787 & -0.2347 \\ 0.0898 & 0.3726 & 0.5493 & 0.2839 \\ -0.0719 & -0.2303 & 0.2834 & 0.8212 \end{bmatrix}$$

We choose the first column from the null-space so that,

$$\mathbf{b} = \begin{bmatrix} 0.1996 \\ 0.7146 \\ 0.2723 \\ -0.5712 \end{bmatrix} + \frac{1}{s} \begin{bmatrix} 0.1143 \\ 0.1512 \\ 0.0898 \\ -0.0719 \end{bmatrix}$$

This choice of $\mathbf{b}$ ensures that the fifth-eighth eigenvalues are retained. Now, using this $\mathbf{b}$ the gains $\mathbf{g}$ and $\mathbf{f}$ are determined in exactly the same way as before,

$$\mathbf{g} = \begin{bmatrix} 10.1513 \\ 12.1105 \\ 8.1401 \\ 7.2688 \end{bmatrix}; \ \mathbf{f} = \begin{bmatrix} 4.4973 \\ 5.4989 \\ 6.4880 \\ 11.2309 \end{bmatrix}$$

3.3 Robust Pole Placement

Mottershead et al. (2009) determined the sensitivity of the poles to small changes in the control gains,

$$\frac{\partial \mu_k}{\partial g_i} = \frac{-\mathbf{e}_i^T \mathbf{N}(\mu_k)\mathbf{b}}{\frac{\partial d}{\partial s}|_{s=\mu_k} + (\mathbf{g} + \mu_k \mathbf{f})^T \frac{\partial \mathbf{N}}{\partial s}|_{s=\mu_k} \mathbf{b} + \mathbf{f}^T \mathbf{N}(\mu_k)\mathbf{b}} \tag{43}$$

$$\frac{\partial \mu_k}{\partial f_i} = \frac{-\mu_j \mathbf{e}_i^T \mathbf{N}(\mu_k)\mathbf{b}}{\frac{\partial d}{\partial s}|_{s=\mu_k} + (\mathbf{g} + \mu_k \mathbf{f})^T \frac{\partial \mathbf{N}}{\partial s}|_{s=\mu_k} \mathbf{b} + \mathbf{f}^T \mathbf{N}(\mu_k)\mathbf{b}} \tag{44}$$

and the sensitivity of the poles to small measurement errors was given by Tehrani et al. (2010),

$$\frac{\partial \mu_k}{\partial h_{pq}} = \frac{-(\mathbf{g} + \mu_k \mathbf{f})^T \mathbf{e}_p \mathbf{e}_q^T \mathbf{b}}{(\mathbf{g} + \mu_k \mathbf{f})^T \left(\frac{\partial \mathbf{H}}{\partial s}|_{s=\mu_k} - \mathbf{e}_p \frac{\partial H_{pq}}{\partial s}|_{s=\mu_k} \mathbf{e}_q^T \right) \mathbf{b} + \mathbf{f}^T \mathbf{H}(\mu_k)\mathbf{b}} \tag{45}$$

We see from equations (43) and (44) that the eigenvalues are insensitive to the gains when $\mathbf{N}(\mu_k)\mathbf{b} = 0$, which is once again the uncontrollability condition.

For robust pole placement the sensitivity-matrix terms should be minimised,

$$\mathbf{S} = \begin{bmatrix} \vdots & \vdots & \vdots & \vdots & \vdots & \vdots & \vdots \\ \cdots & \frac{\partial \mu_k}{\partial g_i} & \cdots & \frac{\partial \mu_k}{\partial f_i} & \cdots & \frac{\partial \mu_k}{\partial h_{pq}} & \cdots \\ \vdots & \vdots & \vdots & \vdots & \vdots & \vdots & \vdots \end{bmatrix} \tag{46}$$

3.4 Practical Aspects

In this section we consider the practical application of active pole placement using electromagnetic inertial actuators and accelememeters. The implementation is achieved by using the dSPACE system with MATLAB-Simulink. In the theoretical sections we considered displacement and velocity feedback. However, the vibration response is measured using accelerometers and in order to obtain the displacement response double integration must be carried out digitally using dSPACE. Digital processing introduces control delays and errors. Therefore, instead of displacement and velocity feedback we now use velocity and acceleration, thereby eliminating one integration and reducing the integration errors. The acceleration/voltage transfer function for such an actuator was given by Preumont (2002),

$$\frac{a(s)}{V(s)} = \frac{c_1 s^2}{m_p s^2 + (C_p + c_1 c_2)\, s + K_p} \tag{47}$$

where m_p is the reaction mass and C_p and K_p denote the damping and stiffness of the support attaching the reaction mass to the actuator casing. The terms c_1 and c_2 are electromagnetic constants. We see that use of the inertial actuator introduces an additional pole for each actuator used.

One very significant advantage of the receptance approach is that the use of an actuator model, such as equation(47), becomes unnecessary if the actuator is included in the measured receptance. Strictly the receptance is a displacement/force transfer function, so from now on we use the term 'receptance' very loosely as in output/input transfer function, in this case specifically as acceleration/voltage. Of course the accelerometers (which may also have significant dynamics, may also be included within this receptance. One of the problems of using an inertial actuator is that the actuator poles may be destabilised as a result of spillover. Therefore it is necessary to place a sensor on the reaction mass to form a nested feedback loop as shown in Figure 7. The reaction mass is not accessible in all inertial actuators. In the experiment described in this section we used the Data Physics IV40 actuator.

The experiment consists of a short cantilever beam with added mass at the tip sufficient to cause the node of the second mode to move to the tip, where the actuator is located as shown in Figure 8. Two accelerometers were used, one at the tip of the beam and the other on the actuator reaction mass. Since the actuator is placed at the node, the second mode is uncontrollable and remain unaffected by control action. This is an example of partial pole placement. Our objective is to place the actuator pole and the first pole of the beam while the second pole of the beam remains unchanged.

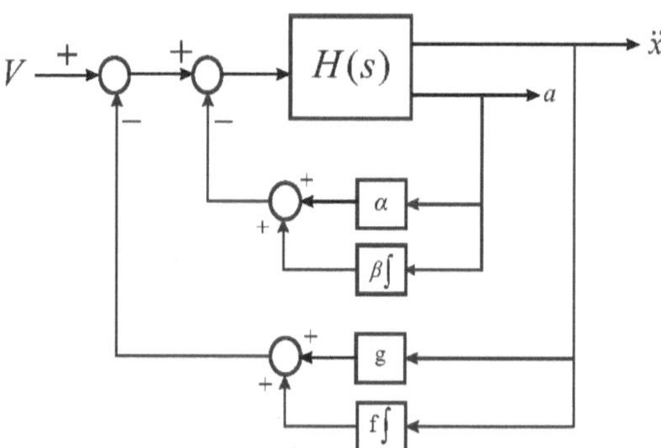

Figure 7: Nested controller

The open-loop poles for the beam and actuator are,

Figure 8: Beam experiment

$\lambda_{b1,2} = -5 \pm 419\mathtt{i}$
$\lambda_{a1,2} = -8 \pm 205\mathtt{i}$

The closed-loop prescribed values are,

$\mu_{b1,2} = -8 \pm 440\mathtt{i}$
$\mu_{a1,2} = -12 \pm 170\mathtt{i}$

Four equations of the following form must be solved,

$$1 + \left(\mu_k^2 \mathbf{g} + \mu_k \mathbf{f}\right)^T \mathbf{H}\left(\mu_k\right) \mathbf{b} = 0; \; k = 1:4$$

$$\mathbf{g} = \begin{pmatrix} g \\ \alpha \end{pmatrix}; \;\; \mathbf{f} = \begin{pmatrix} f \\ \beta \end{pmatrix}$$

Receptance $\mathbf{H}(s)$ is obtained from the measured $\mathbf{H}(\mathtt{i}\omega)$, itself obtained by curve-fitting using the PolyMAX routine (Peeters et al., 2004),

$$\mathbf{H}(\mathtt{i}\omega) = \sum_{j=1}^{n} \left(\frac{\varphi_j \varphi_j^T}{\mathtt{i}\omega - \lambda_j} + \frac{\varphi_j^* \varphi_j^{*T}}{\mathtt{i}\omega - \lambda_j^*} \right) - \frac{LR}{\omega^2} + UR \tag{48}$$

where LR and UR are the upper and lower residuals representing the out of range modes. The measured and curve-fitted receptances are shown in Figure 9. The control gains are found to be,

$$\mathbf{g} = \begin{pmatrix} -2.9794 \\ 0.1508 \end{pmatrix}; \;\; \mathbf{f} = \begin{pmatrix} -36.2843 \\ 18.5617 \end{pmatrix}$$

The simulated open-loop and closed-loop displacement/voltage receptances are shown in Figure 10 where it can be seen that the closed-loop peaks correspond closely to the imaginary parts of the assigned eigenvalues. Figure 11 shows the measured actuator receptance in a full experimental implementation. The same actuator was used to provide the modal-test input as well as producing the control force. A programme of further experimental research is presently being undertaken and will be reported in due course.

4 Conclusions

The receptance approach to pole placement in passive structural modification and active vibration control has been described and explained. The method is based entirely upon data from standard vibration tests and does not depend upon known or evaluated system matrices **M**, **C**, **K**. The system equations are expressed in the units of displacement, whereas force equations are formed when using the conventional **M**, **C**, **K** method. This eliminates the need for model reduction and the use of an observer to estimate unmeasured states. Active control offers much greater flexibility than passive modification in pole assignment. In principle, all $2n$ poles may be assigned actively using a single actuator. A distinct advantage in practical implementation is that elements of the control system, typically actuators, filters and sensors, which must be modelled by the conventional matrix method, may be included in the receptance measurement.

Bibliography

A. A. Berman and W. G. Flannelly. Theory of incomplete models of dynamic structures. *AIAA Journal*, 9:1481–1487, 1971.

B. N. Datta, S. Elhay, and Y. M. Ram. Orthogonality and partial pole placement for the symmetric definite quadratic pencil. *Linear Algebra and its Applications*, 257:366–384, 1997.

W. J. Duncan. The admittance method for obtaining the natural frequencies of systems. *Philosophical Magazine*, 32:401–409, 1941.

I. Fawzy and R. E. D. Bishop. On the dynamics of linear non-conservative systems. *Proc. Royal Society of London*, 352:25–40, 1976.

J. Kautsky, N. K. Nichols, and P. Van Dooren. Robust pole assignment. *International Journal of Control*, 41:1129–1155, 1985.

A. Kyprianou, J. E. Mottershead, and H. Ouyang. Structural modification. part 2: assignment of natural frequencies and antiresonances by an added beam. *Journal of Sound and Vibration*, 284:267–281, 2005.

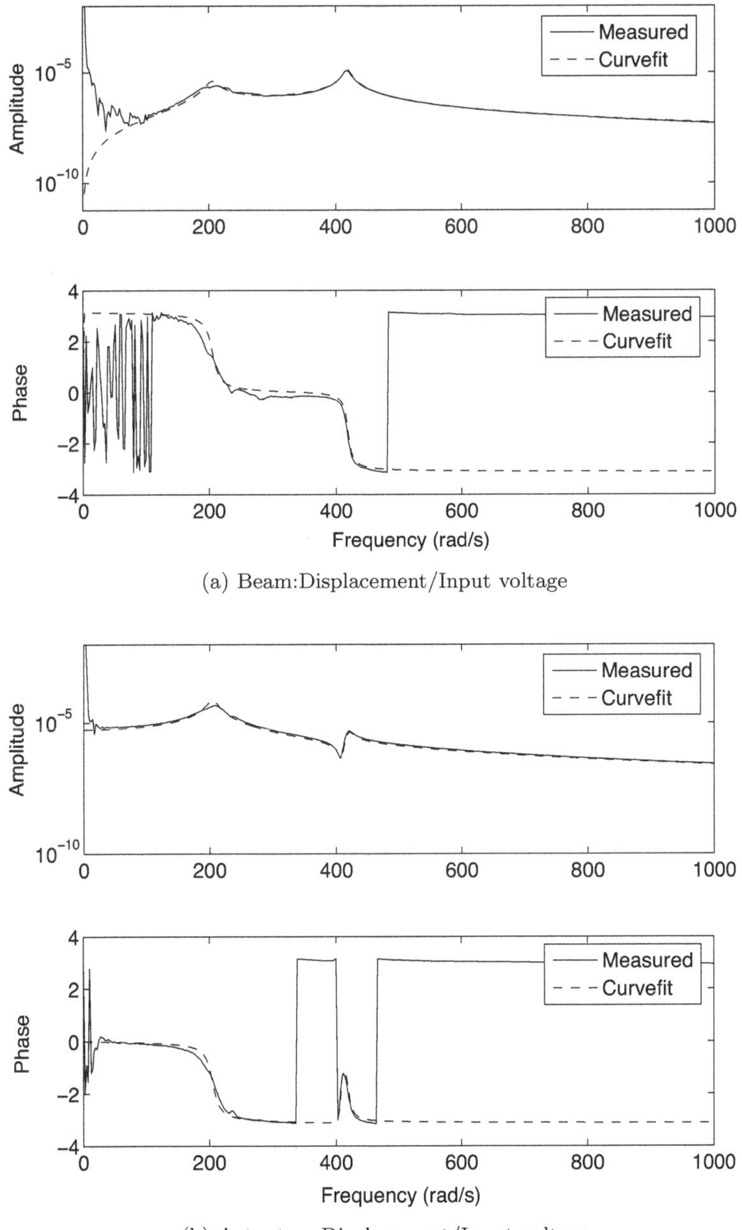

(a) Beam:Displacement/Input voltage

(b) Actuator: Displacement/Input voltage

Figure 9: Measured and curvefitted receptances

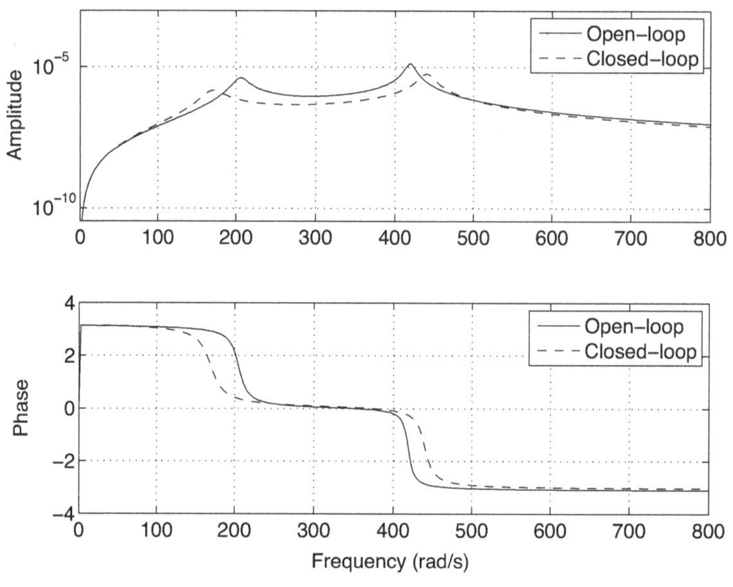

(a) Beam:Displacement/Input voltage

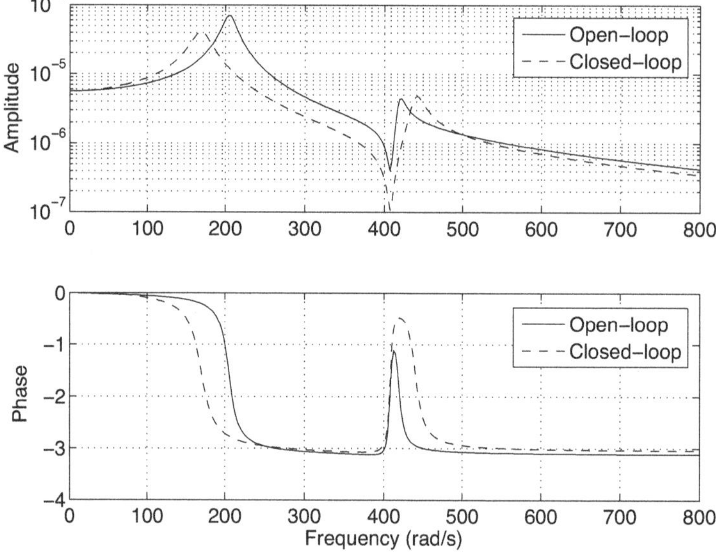

(b) Actuator:Displacement/Input voltage

Figure 10: Simulated pole-placement results

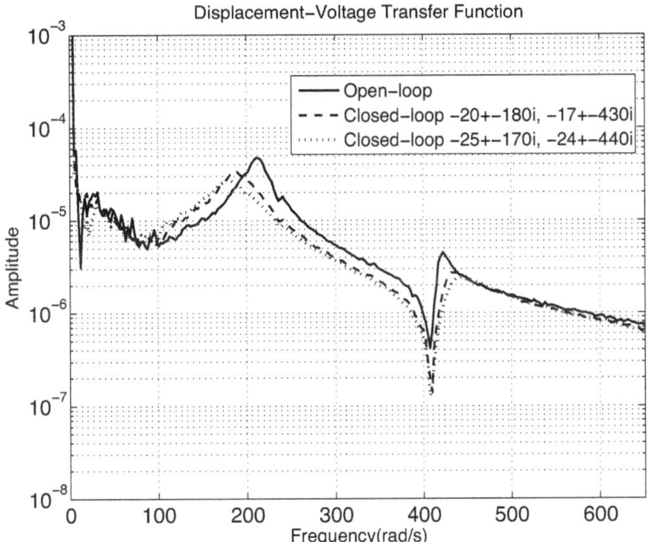

Figure 11: Experimental implementation of pole placement for the beam experiment

J. E. Mottershead and Y. M. Ram. Inverse eigenvalue problems in vibration absorption: passive modification and active control. *Mechanical Systems and Signal Processing*, 20:5–44, 2006.

J. E. Mottershead, C. Mares, and M. I. Friswell. An inverse method for the assignment of vibration nodes. *Mechanical Systems and Signal Processing*, 15:87–100, 2001.

J. E. Mottershead, A. Kyprianou, and H. Ouyang. Structural modification. part 1: rotational receptances. *Journal of Sound and Vibration*, 284: 249–265, 2005.

J. E. Mottershead, M. G. Tehrani, D. Stancioiu, and H Shahverdi. Structural modification of a helicopter tailcone. *Journal of Sound and Vibration*, 289:366–384, 2006.

J. E. Mottershead, M. G. Tehrani, S. James, and Y. M. Ram. Active vibration suppression by pole-zero placement using measured receptances. *Journal of Sound and Vibration*, 311:1391–1408, 2008.

J. E. Mottershead, M. G. Tehrani, and Y. M. Ram. Assignment of eigenvalue sensitivities from receptance measurements. *Mechanical Systems and Signal Processing*, 23:1931–1939, 2009.

B. Peeters, H. Van der Auweraer, P. Guillaume, and J. Leuridan. The polymax frequency-domain method: a new standard for modal parameter estimation? *Shock and Vibration*, 11:395–409, 2004.

A. Preumont. *Vibration Control of Active Structures, Second Edition.* Kluwer Academic Publishers, 2002.

Y. M. Ram and J. E. Mottershead. Receptance method in active vibration control. *American Institute of Aeronautics and Astronautics Journal*, 45:562–567, 2007.

Y. M. Ram, A. Singh, and J. E. Mottershead. State feedback with time delay. *Mechanical Systems and Signal Processing*, 23:1940–1945, 2009.

M. G. Tehrani, W. Wang, C. Mares, and J. E. Mottershead. The generalized Vincent circle in vibration suppression. *Journal of Sound and Vibration*, 292:661–675, 2006.

M. G. Tehrani, J. E. Mottershead, A. T. Shenton, and Y. M. Ram. Robust pole placement in structures by the method of receptances. *Mechanical Systems and Signal Processing*, in press:1–11, 2010.

Solution of Inverse Problems in Biomechanical Imaging

Assad A Oberai [*] and Paul E Barbone[†]

[*] Mechanical Aerospace and Nuclear Engineering, Rensselaer Polytechnic Institute, Troy, NY, USA

[†] Department of Mechanical Engineering, Boston University, Boston, MA, USA

Abstract In this set of notes we consider inverse problems in biomechanical imaging. We briefly describe the clinical relevance of these problems and how the measured data is acquired. We focus on two distinct strategies for solving these problems. These include a direct approach that is fast but relies on the availability of complete interior displacement measurements. The other is an iterative approach that is computationally intensive but is able to handle incomplete interior data and is robust to noise. We present some recent results obtained using these approaches and also comment on future directions of research.

1 Introduction

Certain types of diseases lead to changes in the microstructural organization of tissue. Altered microstructure in turn leads to altered macroscopic tissue properties, which are often easier to infer and quantify than microscopic properties. Thus the measurement of macroscopic properties offers a window into tissue microstructure and health. Consider, for example, the progression of breast cancer during which cancerous cells in the milk duct activate fibroblasts which are responsible for enhanced recruitment of collagen (Kass et al., 2007). At the macroscopic level the enhanced collagen manifests itself in the form of a tumor with increased stiffness. By palpating the tissue in and around the tumor, one may then discern that it is stiff and hence infer that it may be cancerous. Similar associations may be developed for the cirrhosis of liver and for atherosclerosis in arteries. In biomechanical imaging (BMI) we aim to utilize this association between macroscopic mechanical properties of tissue and its health by generating images of the mechanical properties and using these to infer tissue pathology.

Broadly speaking BMI involves the following steps:

1. The tissue is continuously imaged using a standard (phase sensitive) imaging technique like ultrasound or MRI.
2. It is deformed while it is being imaged. The deformation may be applied externally or it may be produced internally (by the pressure changes during a cardiac cycle, for example). Further, it may have a desired time dependence (quasi-static, time-harmonic, impulsive, etc.) and a desired amplitude (infinitesimal strain versus finite strain).
3. The images of the tissue during its deformation are used to determine the displacement field within the tissue. For example, when ultrasound is used to image quasi-static deformation of tissue, successive ultrasound images are registered using cross-correlation in order to determine the intervening displacement field.
4. Using the displacement field in conjunction with an appropriate mechanical model, an inverse problem is solved to determine the spatial distribution of the material parameters.

This set of notes is concerned with the last step in BMI. That is solving the inverse problem of determining the spatial distribution of mechanical properties from the knowledge of its displacements.

1.1 Types of Inverse Problems in BMI

The inverse problems encountered in BMI may be classified on the following basis.

The time-dependence of the excitation. The excitation may be slow so that the response is *quasi-static*. It may also be *time-harmonic, or impulsive, or periodic* as induced by the cardiac cycle.

The magnitude of the induced strain. The excitation may induce small strain (less than 2 %) so that *infinitesimal strain* elasticity is appropriate, or it may induce large strains so that *finite strains* have to be modeled.

The amount of data. In some instances it is possible to measure all components of the displacement field everywhere within the imaging volume of the tissue. For example, when using MRI in conjunction with a time-harmonic excitation all components of the displacement field can be measured. In contrast to this in some instances only select components of the displacement fields are determined with reasonable accuracy. For example when ultrasound is used to determine the displacement field in a quasi-static deformation, the component of displacement along the direction of propagation of the ultrasound waves (called the axial direction) is known much more accurately than the component perpendicular to this direction (called the lateral direction). Sometimes the lateral displacement estimates are so noisy that they are discarded while solving the inverse problem. Thus we may have inverse problems with *complete* or *incomplete* interior displacement data.

From the point of view of developing algorithms for solving inverse problems in BMI, the final classification, that is the one based on the amount of data, is the most relevant. This is because when complete interior data is available one may interpret the equations for the balance of linear momentum of the tissue as partial differential equations (PDEs) for the unknown material parameters. Thus the solution of the inverse problem is reduced to developing methods for solving these PDEs and one is lead to the so-called direct methods which do not involve an iterative solution strategy. This approach fails with incomplete interior data because in this case some terms in these PDEs are not known and iterative methods are employed. When compared with direct methods, iterative methods tend to be more computationally intensive. On the other hand they tend to be more robust, because unlike the direct methods they do not work directly with the noisy measured data and its derivatives. Thus the emphasis in developing new iterative methods is to reduce their computational costs, while the emphasis in developing new direct methods is to enhance their stability and robustness.

In the following sections we describe some of our recent work in developing direct and iterative methods for solving inverse problems in BMI. Several researchers have worked on both these types of methods (Raghavan and Yagle, 1994; Kallel and Bertrand, 1996; Doyley et al., 2000; Sinkus et al., 2000; Oberai et al., 2003; McLaughlin and Renzi, 2006). The description in the following sections is by no means a review of the state of the field, but rather a description of two approaches that we believe have the potential of working in a clinical setting. For a review of the experimental and computational techniques in BMI the interested reader is referred to Gao et al. (1996); Parker et al. (1996); Ophir et al. (1999); Greenleaf et al. (2003); Parker et al. (2005); Barbone and Oberai (2010).

2 A Direct Method Based on the Adjoint Weighted Equations

The equations of balance of linear momentum for an elastic solid may be written as

$$\nabla \cdot (\sum_{n=1}^{N} \boldsymbol{A}^{(n)} \mu_n) = \boldsymbol{f}, \tag{1}$$

where $\mu_1(x), \cdots, \mu_N(\boldsymbol{x})$ are the N scalar material parameters, $\boldsymbol{A}^{(n)}(\boldsymbol{x})$ are N tensors or vectors obtained from the measured displacements $\boldsymbol{u}$ and $\boldsymbol{f}$ is a scalar or a vector field determined completely from the measured displacement that typically represents the effect of inertia. Note that we view (1) as an equation for the material parameters rather than an equation for $\boldsymbol{u}$.

A large number of linear elastic systems may be written in the form of (1). For example:

1. Compressible linear isotropic elasticity: here $N = 2$, and the $\boldsymbol{A}^{(n)}(\boldsymbol{x})$ are tensors. The material parameter $\mu_1 = \mu$ is the shear modulus and $\mu_2 = \lambda$ is the (first) Lamé parameter. Further $\boldsymbol{A}^{(1)} = (\nabla \boldsymbol{u} + \nabla \boldsymbol{u}^T)/2$ and $\boldsymbol{A}^{(2)} = (\nabla \cdot \boldsymbol{u})\mathbf{1}$, where $\boldsymbol{u}(\boldsymbol{x})$ is the measured displacement field, and $\mathbf{1}$ is the identity tensor.
2. Incompressible linear isotropic elasticity: here $N = 2$, and the $\boldsymbol{A}^{(n)}(\boldsymbol{x})$ are tensors. The material parameter $\mu_1 = \mu$ is the shear modulus and $\mu_2 = -p$ is the pressure. Further $\boldsymbol{A}^{(1)} = (\nabla \boldsymbol{u} + \nabla \boldsymbol{u}^T)/2$ and $\boldsymbol{A}^{(2)} = \mathbf{1}$. We note that here we treat the pressure p as an unknown even though it is not a material parameter.
3. Incompressible plane-stress linear isotropic elasticity: Under the plane stress hypothesis the pressure may be determined directly from μ and $\boldsymbol{u}$. As a result the only unknown parameter is μ. Which implies $N = 1$, $\mu_1 = \mu$, and $\boldsymbol{A}^{(1)} = \nabla \boldsymbol{u} + \nabla \boldsymbol{u}^T + 2(\nabla \cdot \boldsymbol{u})\mathbf{1}$.
4. Scalar Helmholtz equation: The case of time-harmonic incompressible linear elasticity is the same as case (2) above where both μ_n and $\boldsymbol{A}^{(n)}$ are allowed to take on complex values. In addition $\boldsymbol{f} = -\rho\omega^2\boldsymbol{u}$, where the density ρ and the frequency ω are assumed to be known. In cases of antiplane deformation, an approximation is commonly made where the pressure term is assumed to be small and ignored. In this case a single component of $\boldsymbol{u}$ (the out of plane component denoted by u) is measured and used to solve for μ. For this case we are lead to the scalar Helmholtz equation with $N = 1$, $\mu_1 = \mu$, and $\boldsymbol{A}^{(1)} = \nabla u$ is a vector.

The system of equations for μ_n implied by (1) has several interesting features not found in typical partial differential equations. They are typically hyperbolic, however, they do not come with any prescribed boundary data. This is because while imaging tissue *in vivo* it is highly unlikely that the material properties on any surface will be known. Also the system of equations are rarely square. Often, there are more equations than unknowns in order to compensate for the lack of boundary data. For these reasons the solution of these equations using a numerical method, such as the finite element method, is challenging. An obvious solution to this problem is to consider using a least-squares approach. However we have found that this approach yields solutions that tend to smooth out sharp transitions in material properties and, worse, to produce errors that propagate far away from the these transitions into the rest of the domain (see Figure 2 for example).

2.1 Adjoint Weighted Equations (AWE)

In order to address the difficulties described above we have developed an alternative approach, that we refer to as the Adjoint Weighted Equations (AWE) (see Barbone et al. (2007); Albocher et al. (2009); Barbone et al. (2010)). These equations are motivated by stabilized finite element methods, when the stabilization parameter tends to infinity. In addition to addressing the difficulties described above, these equations also lead to easily verifiable conditions on measured data that lead to well-posed problems.

The AWE are given by: find $\mu_n \in \mathcal{S}_n$, such that $\forall w_m \in \mathcal{S}_m$,

$$b(w_m, \boldsymbol{\mu}) = l(w_m). \tag{2}$$

Where the bilinear form $b(\cdot,\cdot)$ and the linear form $l(\cdot)$ are given by.

$$b(w_m, \boldsymbol{\mu}) = (\boldsymbol{A}^{(m)} \cdot \nabla w_m, \nabla \cdot (\sum_{n=1}^{N} \boldsymbol{A}^{(n)} \mu_n)), \tag{3}$$

$$l(w_m) = (\boldsymbol{A}^{(m)} \cdot \nabla w_m, f). \tag{4}$$

In the equations above $(\boldsymbol{w}, \boldsymbol{v}) \equiv \int_\Omega \boldsymbol{w}^* \cdot \boldsymbol{v} \, d\boldsymbol{x}$ represents the L_2 inner product for complex-valued functions $\boldsymbol{w}$ and $\boldsymbol{v}$.

In many instances more than one measurement is made in order to render the problem unique. For example, the tissue may be compressed during one measurement and sheared during another. Another alternative is to compress the tissue non-uniformly. Yet, another alternative is to use waves that propagate in different directions. In all these cases there are M equations of the type of (1) that are satisfied. This is easily accommodated within the AWE formulation by simply adding up the bilinear forms corresponding to each deformation. The premise here is that all bilinear forms are positive semi-definite, thus adding them produces a composite form that is "more" positive. The AWE for multiple measurements are also given by (2) where,

$$b(w_m, \boldsymbol{\mu}) = \sum_{l=1}^{M} (\boldsymbol{A}^{(m,l)} \cdot \nabla w_m, \nabla \cdot (\sum_{n=1}^{N} \boldsymbol{A}^{(n,l)} \mu_n)), \tag{5}$$

$$l(w_m) = \sum_{l=1}^{M} (\boldsymbol{A}^{(m,l)} \cdot \nabla w_m, f^{(l)}). \tag{6}$$

where $\boldsymbol{A}^{(\cdot,l)}$ are the tensors corresponding to the lth measurement, and $\boldsymbol{f}^{(l)}$ are the corresponding vectors.

A viable numerical method is constructed by the Galerkin discretization of the AWE formulation using standard finite element function spaces. This method is given by: find $\mu_n^h \approx \mu_n \in \mathcal{S}_n^h \subset \mathcal{S}_n$ be such that $\forall w_m^h \in \mathcal{S}_m^h \subset \mathcal{S}_m$,

$$b(w_m^h, \boldsymbol{\mu}^h) = l(w_m^h). \tag{7}$$

2.2 Properties of the AWE formulation

In order to prove the well-posedness of the continuous AWE and optimal rates of convergence for their discrete counterpart, the following assumptions are made:

1. The bilinear form defines a natural norm. That is $\forall w_m \in \mathcal{S}_m$, with $\boldsymbol{w} \neq \boldsymbol{0}$,

$$\|\boldsymbol{w}\|_A^2 \equiv \sum_{m=1}^{N} \sum_{n=1}^{N} \sum_{l=1}^{M} (\boldsymbol{A}^{(m,l)} \nabla w_m, \boldsymbol{A}^{(n,l)} \nabla w_n) > 0. \tag{8}$$

2. Generalized Poincare inequality: There exists a positive constant $C_P < \infty$ such that $\forall w_m \in \mathcal{S}_m$,

$$\sum_{m=1}^{N} \sum_{n=1}^{N} \sum_{l=1}^{M} (w_m \nabla \cdot \boldsymbol{A}^{(m,l)}, w_n \nabla \cdot \boldsymbol{A}^{(n,l)}) \leq C \|\boldsymbol{w}\|_A^2, \forall C \geq C_P. \tag{9}$$

3. Boundedness of the A-norm. That is there exists a finite, positive constant so that $\forall w_m \in \mathcal{S}_m$,

$$\|\boldsymbol{w}\|_A \leq C \|\boldsymbol{w}\|_1, \tag{10}$$

where $\|\cdot\|_1$ denotes the H^1 norm.

Under these conditions we are able to prove that the continuous solution to the AWE, denoted by $\boldsymbol{\mu}$, exists and is unique. Further for the Galerkin solution $\boldsymbol{\mu}^h$, we have the following optimal error estimate:

$$\|\boldsymbol{\mu} - \boldsymbol{\mu}^h\|_A \leq Ch^p, \tag{11}$$

where h denotes the mesh size and p is order of the complete polynomial represented by the finite element function spaces.

We now present some examples that illustrate the performance of the AWE based finite element method.

2.3 Numerical Examples

Convergence : As the first example we consider a compressible material in plane strain configuration. This is represented by case 1 in Section 2. The material parameters are the Lame parameters μ and λ, and thus in (1) $N = 2$. In order to achieve a unique solution two independent displacement fields are required. Corresponding to these we select the strain tensors

$$\boldsymbol{\epsilon}^{(1)} = \begin{bmatrix} 1+x+y & -2(x+y+1) \\ -2(x+y+1) & 1+x+y \end{bmatrix} \tag{12}$$

$$\boldsymbol{\epsilon}^{(2)} = \begin{bmatrix} 7+x+2y & -2(x+2y+5) \\ -2(x+2y+5) & 3+x+2y \end{bmatrix} \tag{13}$$

These yield the exact solution $\mu = \lambda = e^{x+y}$. We solve the problem numerically on grids ranging from 2×2 elements to 256×256 using the AWE method with bilinear finite elements. We compute both the L_2 norm and H^1 semi-norm errors and plot them in Figure 1. For both modulus fields we observe optimal convergence in the L_2 error and better than optimal (super-convergence) convergence in the H^1 semi-norm.

Effect of Discontinuity : Next we consider an incompressible plane stress problem to test the ability of the AWE method to reproduce steep gradients in material properties. This is represented by case 3 in Section 2. In this case the shear modulus μ is the only material parameter, thus $N = 1$. Further, a single measurement is sufficient to uniquely recover μ up to a multiplicative parameter. The exact distribution comprises of a discontinuous circular inclusion embedded in a homogeneous background. The contrast between the inclusion and the background is 5:1 and the problem is posed on a unit square. The "measured" displacement field is obtained by solving the forward problem on a 100×100 mesh with the boundary conditions shown in Figure 2. Thereafter the displacement field is down-sampled to a

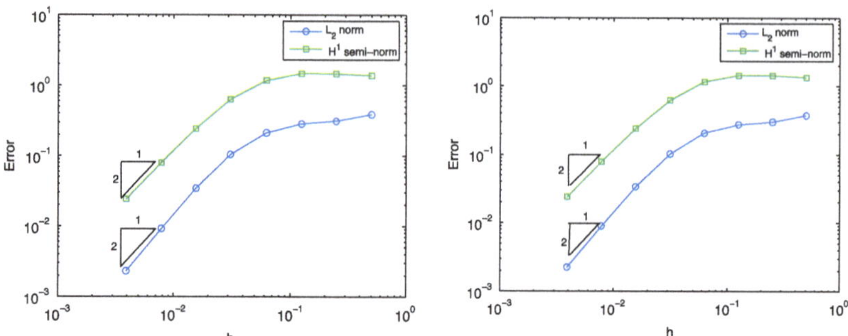

Figure 1. Error as a function of mesh size h in computing the Lamé parameters μ (left) and λ (right).

50×50 mesh and used to extract strain by employing an L_2 projection of the derivatives on to standard finite element basis functions. These strains are then used to reconstruct μ using both the AWE and least squares (LS) formulations with standard bilinear finite elements. The results are shown in Figure 2. From this figure we observe that the AWE solution recovers the shape as well as the steepness of the inclusion well. From the line plot along the diagonal we observe that it recovers the contrast quite well too, while the LS solution has significant errors throughout the domain.

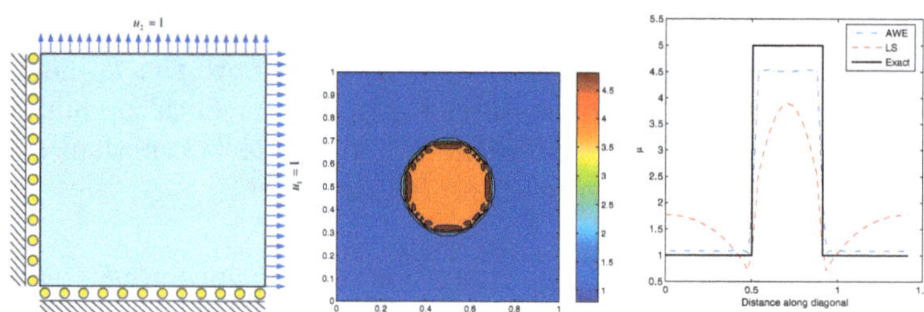

Figure 2. Plane stress problem with a discontinuous distribution. Left: prescribed boundary conditions on the displacement. Center: Contour plot of the reconstructed shear modulus using AWE. Right: Reconstructed shear modulus along the diagonal for AWE and least-squares (LS) methods.

Effect of Noise : Next we consider another incompressible plane stress problem with a circular inclusion. The set up is exactly the same as above except the contrast is diminished (3:1). We generate the "measured" data as in the example above, but now consider three scenarios: with 0%, 3% and 10% additive white Gaussian noise in strains. The results are shown in Figure 3. We observe that the reconstruction even with no noise is not perfect. This is likely because the assumptions for the optimal convergence of μ^h are violated for this data. As we add noise the reconstructions get worse. However even with 10% we are able to discern the inclusion and obtain a rough estimate of the contrast. It is to be noted that this is achieved with no regularization.

Experimental Data : In the final example we consider the reconstruction of shear modulus using experimental data. In this case the material is made from gelatin and the modulus distribution comprises of a stiff cylindrical inclusion (made with a higher concentration of gelatin) within a softer homogeneous background. The diameter of the inclusion is about one-half the horizontal extent of the phantom and it is centered about 2/3rd of the way up from the bottom edge. The excitation is in the form a radiation force applied to one side of the inclusion. This generates a wave that travels from left to right and traverses the inclusion. Another excitation is made at the other side of the inclusion leading to a wave the travels from right to left. The time dependent out-of-plane component of the displacement corresponding to these excitations is measured by cross-correlating successive ultrasound images. In processing this data the main dilatational wave is filtered out by recognizing that it travels much faster than the shear wave. This measurement as well as the techniques used to extract displacement data from it are described in Parker (2010).

The displacement data are Fourier-transformed in time and the scalar Helmholtz equation is used as a model for the out-of-plane displacement component. This corresponds to case 4 in the discussion in Section 2. As a result in (1), $N = 1$ and $M = 2$. The force term is non-zero and is given by $\rho\omega^2 u$, where the density $\rho = 1000 kg/m^3$ and the frequency $\omega = 2\pi * 333\ rad/s$. The AWE method is used in conjunction with a total variation regularization to solve for the shear modulus at this frequency. Both real and imaginary components of the shear modulus are calculated. In Figure 4, we plot the real component of the shear modulus distribution. In this picture we can clearly see the inclusion, though its shape is somewhat distorted. We observe that the value of the shear modulus is quite close to the experimentally measured value of $10^4 Pa$.

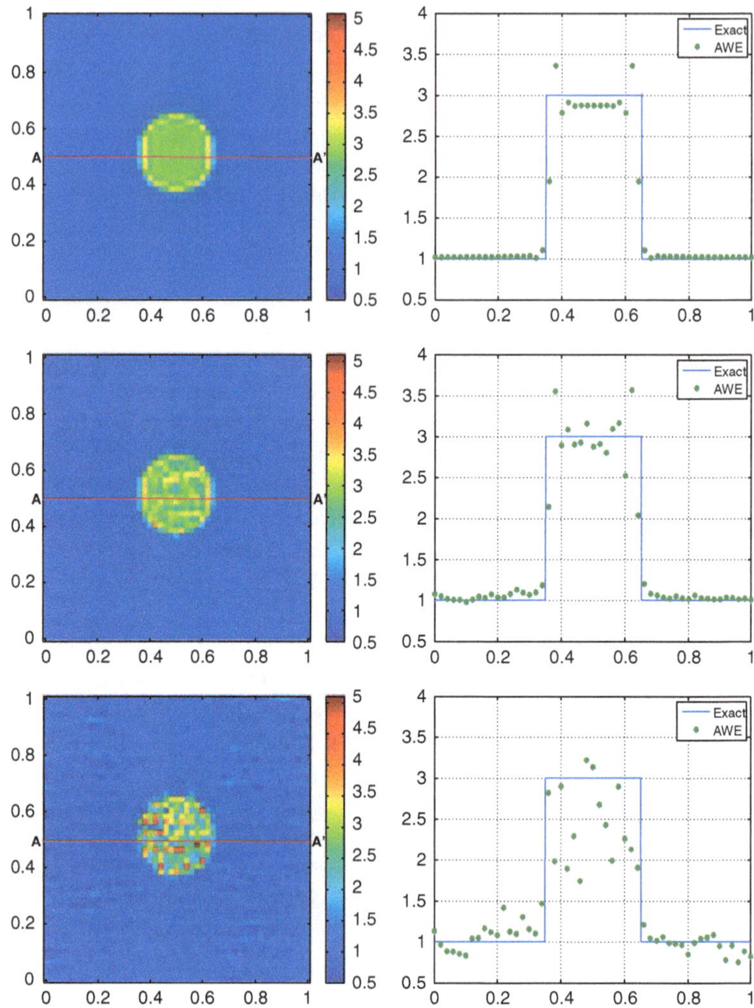

Figure 3. Reconstructions for the plane stress problem in the presence of noise. Top row: 0% added noise. Center row: 3% added noise. Bottom row: 10% added noise. Left column: contour map of shear modulus reconstructed using AWE. Right column: Reconstructed shear modulus along line $A - A'$.

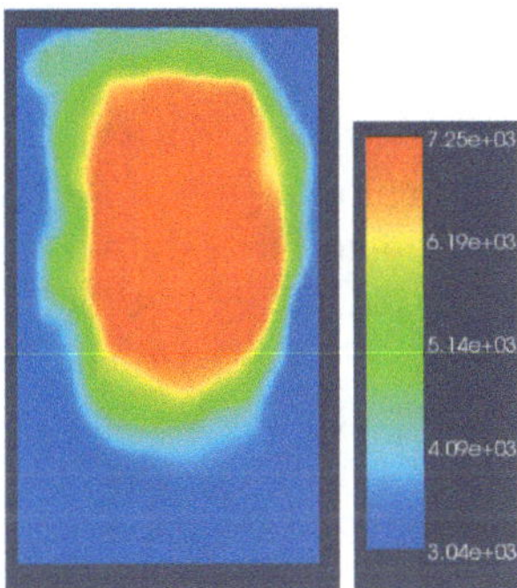

Figure 4. Shear modulus reconstruction using time-harmonic excitation. The units on the colorbar are Pa.

2.4 Future Work

The AWE formulation has shown promise in initial tests. There are however, several outstanding issues that need to be addressed. These include,

1. We have quantified the performance of the AWE formulation for compressible elasticity and for plane stress incompressible elasticity. It is yet to be tested for incompressible elastic materials in plane strain and in three dimensions. These problems are likely to offer additional challenges. For example, in incompressible plane strain elasticity, even with two independent displacement measurements, the solution space is four dimensional. As a result four additional constraints must be imposed. Our initial results indicate that the final solution depends on the form of these constraints as well as how these constraints are applied. This issue needs to be better understood.

2. When multiple displacement fields are used as input, they may be linearly combined to yield other valid displacement fields within the context of linear elasticity. Within the AWE formulation, different combinations lead to different results. It is expected that the method will benefit from a theory of optimal measurement weighting.

3. While initial results indicate that AWE is robust to noise, in practice it is likely that some form of regularization will be required when dealing with measured data (see for example the last case in the previous section). The role of regularization within the AWE formulation and its effect on the solution is yet to be worked out.

3 A Gradient Based Iterative Method

As mentioned earlier the displacement data obtained in BMI is often incomplete in that only certain components of displacements are determined accurately. In addition, the displacement data is noisy. Thus differentiating it in order to evaluate the strain tensors $\boldsymbol{A}^{(n)}$ is problematic. Motivated by this, several researchers have adopted an iterative strategy where the equations (1) are not solved directly for the material parameters.

In the iterative approach a functional that measures the difference between the measured and a predicted displacement field is minimized. The predicted displacement field is determined by solving the forward elasticity problem with an assumed distribution of elastic parameters. This distribution is improved upon iteratively, until the difference between the predicted and measured displacements is minimal. In the following paragraphs we describe our iterative approach to solving the inverse elasticity problem. We describe this method in the context of solving the nonlinear inverse elasticity problem with quasi-static deformation. The interested reader is referred to Oberai et al. (2003); Gokhale et al. (2008); Oberai et al. (2009) for a detailed description.

The functional to be minimized is of the form

$$\pi[\boldsymbol{\mu}] = \sum_{l=1}^{M} \frac{1}{2} \| \boldsymbol{T}(\boldsymbol{u}_{meas}^{(l)}) - \boldsymbol{T}(\boldsymbol{u}^{(l)}) \|^2 + \sum_{n=1}^{N} \frac{\alpha_n}{2} \mathcal{R}[\mu_n]. \tag{14}$$

In the equation above $\boldsymbol{u}_{meas}^{(l)}$ are the measured displacement fields corresponding to M different excitations, and $\boldsymbol{u}^{(l)}$ are the corresponding predicted displacement fields obtained by solving the forward nonlinear elasticity problem with assumed material parameter distributions. Thus these displacement fields are implicit functions of the N material parameters $\boldsymbol{\mu} = [\mu_1, \cdots, \mu_N]$. Further $\boldsymbol{T}$ is a diagonal tensor that weights the contribution from different components of the displacement fields in inverse proportion to the error in their measurement. In particular, if a given component is unknown then the corresponding entry in $\boldsymbol{T}$ is 0. The functional $\mathcal{R}$ is a regularization term that ensures smoothness in the reconstructed parameters and α_n is the regularization parameter for the material parameter μ_n. Typically we employ a regularized version of the total variation diminishing regularization given by

$$\mathcal{R}[\mu] = \int_{\Omega} \sqrt{\beta^2 + |\nabla \mu|^2} d\boldsymbol{x}, \tag{15}$$

where the parameter β is selected to be small. It ensures the differentiability of the regularization function when $\nabla \mu = \mathbf{0}$.

In BMI enough interior data is available, so that local minima tend not to be a problem. As a result one may use gradient based algorithms to solve the minimization problem efficiently. We use the BFGS quasi-Newton method that builds accurate approximations to the Hessian using gradient information. In evaluating the gradient of the functional π with respect to the material parameters, terms of the type $\int_\Omega \delta \boldsymbol{u}^{(l)} \cdot \boldsymbol{v} \, d\boldsymbol{x}$ appear. In these terms $\boldsymbol{v}$ is a vector that is easily determined, and $\delta \boldsymbol{u}^{(l)}$ represents the variation in $\boldsymbol{u}^{(l)}$ corresponding to a variation $\delta \boldsymbol{\mu}$ in the material properties. Instead of explicitly evaluating $\delta \boldsymbol{u}^{(l)}$, which would require as many linearized forward solves solves as the number of degrees of freedom used to represent $\boldsymbol{\mu}$ (often around 10^4), we directly evaluate the integral $\int_\Omega \delta \boldsymbol{u}^{(l)} \cdot \boldsymbol{v} \, d\boldsymbol{x}$ by introducing corresponding adjoint fields. This reduces the number of linearized forward solves to the number of measured displacement fields (seldom larger than 4).

Our iterative algorithm then proceeds as follows:

1. Guess a material parameter distribution $\boldsymbol{\mu}$.
2. Using $\boldsymbol{\mu}$ solve M nonlinear forward elasticity problems to determine the predicted displacements $\boldsymbol{u}^{(l)}$.
3. Using these and the measured displacement fields, solve M linearized forward problems to determine the adjoint fields.
4. Using the predicted and measured displacements and the adjoint fields determine the functional and the gradient.
5. If the minimum is not obtained, obtain a new guess for $\boldsymbol{\mu}$ from BFGS and go to step 2.

In this algorithm the most expensive step is (2), that is the nonlinear forward elasticity solves. In a typical finite deformation elasticity problem each solve involves incrementing the load parameter and performing several Newton iterations to obtain a converged displacement solution at that value of the load parameter. This is continued until the final load configuration is obtained. We have noted that the sequence of the loading steps can be eliminated if one starts from the fully loaded configuration corresponding to the previous guess of the material properties. Thus instead of a continuation in the load parameter, we perform a continuation in material properties. This can be performed at all iterates of the BFGS algorithm except the first, where continuation in the loading parameter has to be used.

Another challenge in solving the finite deformation elasticity problem is the treatment of the incompressibility constraint. Several techniques have been proposed to handle this problem. We have found that most of these require very small load (or material property) increments in order to converge. The best strategy that we have used is a stabilized finite element method proposed by Klaas et al. (1999) that allows for equal order interpolations and is robust.

3.1 Examples

In this section we present two examples that demonstrate the utility of our iterative strategy.

Reconstructions in 3D with Simulated Data We test the performance of our algorithm on simulated data. We create a forward model of a incompressible hyperelastic material described by the strain energy density function in Gokhale et al. (2008), which is a simplified version of the Veronda-Westmann strain energy density function. The strain energy function contains two parameters; μ, corresponds to the shear modulus of the material at small strains, and γ determines the rate of exponential increase in stress with strain and thus accounts for the nonlinear behavior of the material. The sample comprises of a hard inclusion embedded within a softer background. In addition, the inclusion hardens with strain at a faster rate than the background. In particular, the shear modulus values are $\mu_{background} = 1$ and $\mu_{inclusion} = 5$, while the values for the nonlinear parameter are $\gamma_{background} = 1$ and $\gamma_{inclusion} = 5$ (see Figure 5).

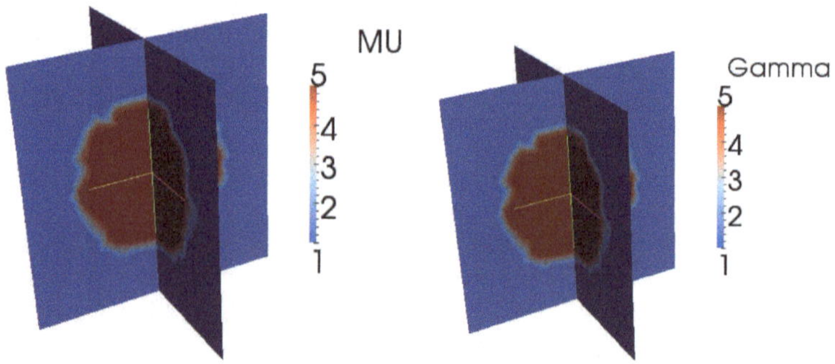

Figure 5. Distribution of the shear modulus μ (left) and the nonlinear parameter γ (right) for the simulated tissue-mimicking phantom.

This specimen was compressed to about 20% strain and 3% Gaussian random noise was added to the displacements. The axial displacement fields at about 1% strain and 20% strain were used in the algorithm to recover the shear modulus and nonlinear parameter distributions. TVD regularization was used to handle the noise in the displacements. The reconstructed distribution of parameters is shown in Figure 6. We observe that for both μ and γ we recover the distribution quite well.

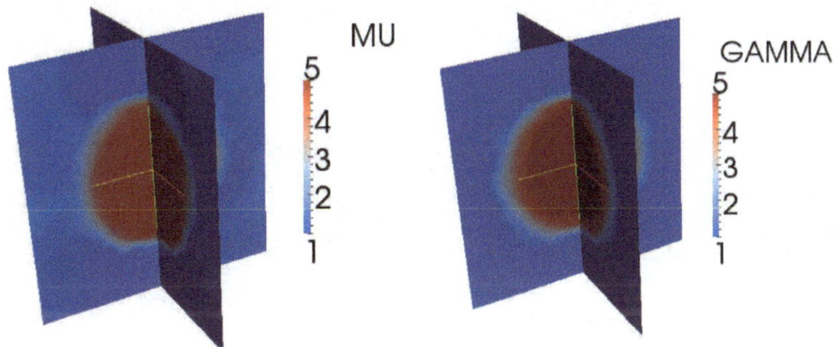

Figure 6. Distribution of the reconstructed shear modulus μ (left) and the nonlinear parameter γ (right) for the simulated tissue-mimicking phantom.

Reconstructions in 2D With Clinical Data We now describe some of our latest *in-vivo* clinical results. Radiofrequency (RF) echo data was acquired for patients with benign and malignant breast lesions. The details of the study as well as the associated compliances are described in Oberai et al. (2009). A free-hand compression technique was used to induce displacements with maximum overall strain in the range of 20%. The displacements were obtained using a guided-search block-matching algorithm described in Jiang and Hall (2007). Displacement estimates were obtained on a $200 \times 200\mu m$ grid and were down-sampled by a factor of four for the modulus reconstructions. Out of the ten specimens included in this study we present the results for one malignant and one benign lesion.

In the modulus reconstruction we assumed plane stress and utilized two deformation fields, one at an overall strain of about 1% and another at a strain of about 12%. We determined the shear modulus distribution up to a multiplicative parameter using the small deformation displacement field and thereafter determined the nonlinear parameter γ up to an additive constant using the large deformation displacement field. We employed the exponential strain energy density function described in Gokhale et al. (2008) and TVD regularization for both μ and γ.

The results for the benign lesion (a fibroadenoma) are shown in Figure 7. We can clearly observe the lesion as a region of elevated stiffness in the image for the shear modulus. We also note that the nonlinear parameter within the lesion is slightly elevated compared to the background. The reconstructions for the malignant lesion (Infiltrating ductal carcinoma) are shown in Figure 8. Once again the lesion clearly stands out in the shear modulus image. In addition, in this case the value of the nonlinear parameter within the lesion is much higher than the background. These results are indicative of the ability of the nonlinear strain hardening parameter in distinguishing different types of stiff breast lesions.

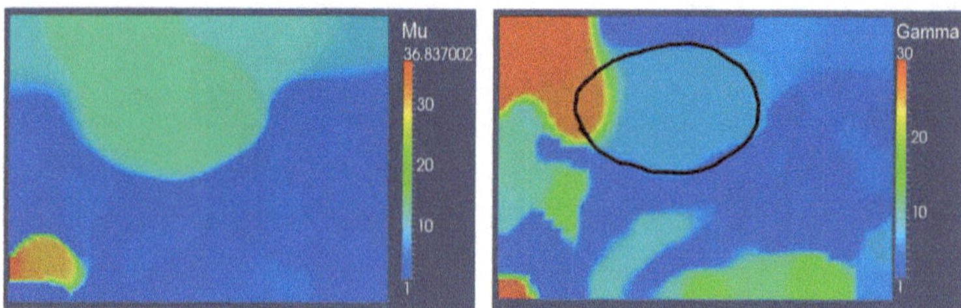

Figure 7. Reconstruction of the shear modulus μ (left) and nonlinear strain hardening parameter γ (right) for a benign lesion.

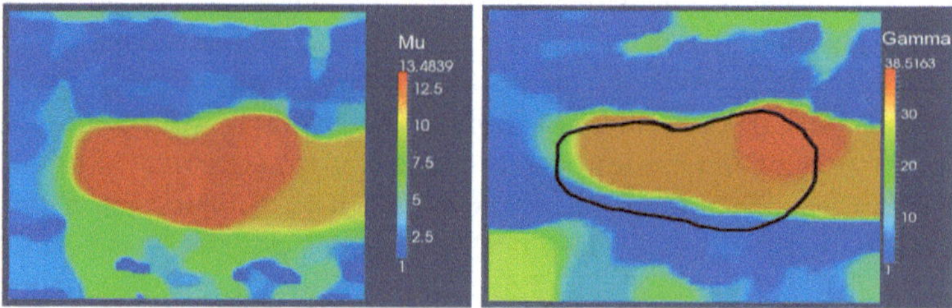

Figure 8. Reconstruction of the shear modulus μ (left) and nonlinear strain hardening parameter γ (right) for a malignant lesion.

3.2 Future Work

With our current approach we are able to solve clinically relevant inverse elasticity problems. However, we are still quite a ways away from our goal of being able to do so in three dimensions and in almost real-time. There are several outstanding issues that are yet to be dealt with including,

1. Selection of appropriate strain energy density function. For now we have used a modified version of the Veronda-Westmann (VW) strain energy density function that reproduces the exponential stress-strain behavior observed in most tissues. The original VW function contains three parameters (Veronda and Westman, 1970). We have reduced these two in order to ameliorate the non-uniqueness of the inverse

problem. One of these represents the shear modulus at small strains and another represents the degree of stiffening of tissue with increasing strain. However recent measurements by O'Hagan and Samani (2009) indicate that the VW function even with three parameters may not be the best strain energy density function for breast tissue. Other functions such as the Yeoh model are better able to represent tissue behavior and need to be considered.

2. Selection of regularization parameter. When using Tikhonov regularization one may use the L-curve or Morozov discrepancy principle in order to estimate the regularization parameter. Our experience suggests that neither of these strategies provide enough discrimination when using TVD regularization. Thus a better strategy to select α_n is needed.

3. Boundary conditions. In order to determine the predicted displacement fields, a well-posed nonlinear forward problem has to be solved. This requires displacement or traction boundary conditions on image edges. An obvious choice would be to use the measured displacements, but this approach leads to strong imposition of noisy displacement data, and is not possible when only one component of the displacement field is known. We have recently devised an approach that circumvents the imposition of BCs for the linear elasticity problem. The extension of this to the nonlinear is yet to be worked out.

4. Uniqueness. The uniqueness of the linear inverse problem in two dimensions in plane-strain and plane-stress configurations has been worked out. We have recently made some progress in extending these results to three dimensions and to the nonlinear problem in two dimensions, where we have considered the Neo-Hookean and modified Veronda-Westmann strain energy density functions. Similar results for other functions that better model soft tissue behavior have not been developed.

5. Computational time. Even with the adjoint approach and the continuation in material parameters, a typical nonlinear inverse problem on a 100×100 grid takes about 20 minutes to solve on a fast quad-core computer. It will be of great clinical importance to reduce this time down to a minute or two, so as to provide feedback to the radiologist while the patient was still in the clinic. There are several strategies that can be applied to reduce this time including parallelization of the linear solves and the matrix assembly procedure, smoothing strategies to accelerate iterative convergence, and multigrid techniques for both material parameter and displacement discretizations.

4 Conclusions

In Biomechanical Imaging (BMI), images of the spatial distribution of mechanical properties of tissue are generated from measurements of its response to an excitation. In generating these images BMI relies on the solution of an inverse problem of determining mechanical properties of a material from measurements of its displacement field(s). This problem differs from other "traditional" inverse problems in that significant interior data is available and should be utilized. New inverse problems techniques that address this issue are therefore necessary to solve it. In this set of notes we have described two such approaches. We have also presented some representative results obtained from these techniques and described avenues of future research.

Acknowledgements

The authors gratefully acknowledge the industry and ingenuity of a number of talented collaborators, including (in alphabetical order) Uri Albocher, Jeffrey C. Bamber, Jean-Francois Dord, Sevan Goenzen, Nachiket Gokhale, Timothy J. Hall, Isaac Harari, Michael Richards, Carlos A. Rivas, and Yixiao Zhang. Financial support through grants NIH-R01CA140271, NIH-R21CA133488, NIH RO1 AG029804 and the United States-Israel Binational Science Foundation (BSF) Grant No. 2004400 is also gratefully acknowledged. AAO also acknowledges support from Humboldt Foundation in preparing this set of notes.

Bibliography

U. Albocher, A.A. Oberai, P.E. Barbone, and I. Harari. Adjoint-weighted equation for inverse problems of incompressible plane-stress elasticity. *Computer Methods in Applied Mechanics and Engineering*, 198(30-32): 2412–2420, 2009.

P.E. Barbone and A.A. Oberai. A Review of the Mathematical and Computational Foundations of Biomechanical Imaging. *Computational Modeling in Biomechanics*, pages 375–408, 2010.

P.E. Barbone, A.A. Oberai, and I. Harari. Adjoint-weighted variational formulation for a direct computational solution of an inverse heat conduction problem. *Inverse Problems*, 23:2325, 2007.

P.E. Barbone, C.E. Rivas, I. Harari, U. Albocher, A.A. Oberai, and Y. Zhang. Adjoint-weighted variational formulation for the direct solution of inverse problems of general linear elasticity with full interior data. *International Journal for Numerical Methods in Engineering*, 81 (13):1713–1736, 2010.

M. M. Doyley, P. M. Meaney, and Bamber J. C. Evaluation of an iterative reconstruction method for quantitative elasticity. *Physics in Medicine and Biology*, 45:1521–1540, 2000.

L. Gao, K.J. Parker, R.M. Lerner, and S.F. Levinson. Imaging of the Elastic Properties of Tissue - A Review. *Ultrasound in Medicine and Biology*, 22(8):959–977, 1996.

N.H. Gokhale, P.E. Barbone, and A.A. Oberai. Solution of the nonlinear elasticity imaging problem. *Inverse Problems*, 24:045010, 2008.

J.F. Greenleaf, M. Fatemi, and M. Insana. SELECTED METHODS FOR IMAGING ELASTIC PROPERTIES OF BIOLOGICAL TISSUES. *Annual Reviews in Biomedical Engineering*, 5(1):57–78, 2003.

Jingfeng Jiang and Timothy J. Hall. A parallelizable real-time ultrasonic speckle tracking algorithm with applications to ultrasonic strain imaging. *Physics in Medicine and Biology*, 52:3773–3790, 2007.

F. Kallel and M. Bertrand. Tissue elasticity reconstruction using linear perturbation method. *IEEE Transactions on Medical Imaging*, 15(3): 299–313, 1996.

L. Kass, J.T. Erler, M. Dembo, and V.M. Weaver. Mammary epithelial cell: influence of extracellular matrix composition and organization during development and tumorigenesis. *International Journal of Biochemistry and Cell Biology*, 39(11):1987–1994, 2007.

O. Klaas, A. Maniatty, and M.S. Shephard. A stabilized mixed finite element method for finite elasticity.:: Formulation for linear displacement and pressure interpolation. *Computer Methods in Applied Mechanics and Engineering*, 180(1-2):65–79, 1999.

J. McLaughlin and D. Renzi. Using level set based inversion of arrival times to recover shear wave speed in transient elastography and supersonic imaging. *Inverse Problems*, 22:707–725, 2006.

A.A. Oberai, N.H. Gokhale, and G.R. Feijoo. Solution of Inverse Problems in Elasticity Imaging Using the Adjoint Method. *Inverse Problems*, 19: 297–313, 2003.

A.A. Oberai, N.H. Gokhale, S. Goenezen, P.E. Barbone, T.J. Hall, A.M. Sommer, and J. Jiang. Linear and nonlinear elasticity imaging of soft tissue in vivo: demonstration of feasibility. *Physics in Medicine and Biology*, 54(5):1191–1207, 2009.

J.J. O'Hagan and A. Samani. Measurement of the hyperelastic properties of 44 pathological ex vivo breast tissue samples. *Physics in Medicine and Biology*, 54:2557, 2009.

J. Ophir, S.K. Alam, B. Garra, F. Kallel, E. Konofagou, T. Krouskop, and T. Varghese. Elastography: ultrasonic estimation and imaging of the elastic properties of tissues . *Proceedings of the Institution of Mechanical Engineers Part H-Journal of Engineering in Medicine*, 213 (H3):203–233, 1999.

K. J. Parker. Private communication, 2010.

K.J. Parker, L. Gao, R.M. Lerner, and S.F. Levinson. Techniques for elastic imaging: a review. *Engineering in Medicine and Biology Magazine, IEEE*, 15(6):52–59, 1996.

K.J. Parker, L.S. Taylor, S. Gracewski, and D.J. Rubens. A unified view of imaging the elastic properties of tissue. *The Journal of the Acoustical Society of America*, 117:2705, 2005.

K.R. Raghavan and A.E. Yagle. Forward and Inverse Problems in Elasticity Imaging of Soft Tissues. *IEEE Transactions on Nuclear Science*, 41: 1639–1648, 1994.

R. Sinkus, J. Lorenzen, J. Schrader, M. Lorenzen, M. Dargatz, and D. Holz. High-resolution Tensor MR Elastography for Breast Tumor Detection. *Physics in Medicine and Biology*, 45:1649–1664, 2000.

D. R. Veronda and R. A. Westman. Mechanical characterization of skin – finite deformations. *Journal of Biomechanics*, 3(1):111–122, January 1970.

Zeitfracht Medien GmbH
Ferdinand-Jühlke-Straße 7
99095 Erfurt, Deutschland
produktsicherheit@kolibri360.de